POLYMER BLENDS

Processing, Morphology,
and Properties
Volume 2

POLYMER BLENDS

Processing, Morphology, and Properties

Volume 2

Edited by

Marian Kryszewski
Andrzej Gałęski

Center of Molecular and Macromolecular Studies
Polish Academy of Sciences
Łódź, Poland

and

Ezio Martuscelli

Institute of Research on Technology of
Polymers and Rheology, CNR
Naples, Italy

PLENUM PRESS • NEW YORK AND LONDON

Library of Congress Cataloging in Publication Data

Joint Italian–Polish Seminar on Multicomponent Polymeric Systems (1st: 1979: Capri, Italy)
Polymer blends.

Vol. 2: Proceedings of the Second Polish –Italian Joint Seminar on Multicomponent Polymeric Systems, held September 7–12, 1982 in Lodz, Poland, edited by Marian Kryszewski, Andrzej Galeski, and Ezio Martuscelli.
Includes bibliographical references and indexes.
1. Polymers and polymerization—Congresses. I. Martuscelli, Ezio. II. Palumbo, Rosario. III. Kryszewski, Marian. IV. Galeski, Andrzej. V. Joint Italian–Polish Seminar on Multicomponent Polymeric Systems (2nd: 1982: Lodz, Poland. VI. Title.
QD380.J64   1979   547.8′4   80-22862
ISBN 0-306-41802-9 (v. 2)

Proceedings of the Second Polish –Italian Joint Seminar on
Multicomponent Polymeric Systems, held September 7–12, 1982,
in Łódź, Poland

© 1984 Plenum Press, New York
A Division of Plenum Publishing Corporation
233 Spring Street, New York, N.Y. 10013

Printed in the United States of America

The study of multicomponent polymeric systems is now one of advanced domains in modern polymer science. Recent years have revealed their growing importance from both scientific and practical points of view.

Various physical properties of polymer blends are particularly interesting as being related to the composition, interaction between components, phase structure as well as to the conditions of processing.

The aim of this volume is to present the recent results discussed at the Second Joint Polish-Italian Seminar on Multicomponent Polymer Systems. The scope of topics to be covered was fairly wide and the participation of scientists from other countries made it possible to discuss some aspects of their studies.

The significance of polymer blends has been an incentive for us to take also into consideration the advances in polymer blend preparation. The general characteristics of multicomponent polymeric systems included the formation and transitions of the complex structure in blends crystalline and amorphous components. Since the interactions between the blend components are of great importance the coupling agent activity and the modification of contacts between the components as well as general aspects of adhesion between polymers have been examined.

The preparation and properties of interpenetrating polymer networks, one of the newest class of materials, have been analysed and compared with other multicomponent polymeric systems.

The description of properties of polymer blends seems to be a major problem which needed consideration, and therefore, this issue has been carefully discussed with special attention to mechanical and electrical properties.

In view of the association that exists between polymer blends and filled materials with regard to phase adhesion and phase

structure, several appropriate problems have been considered
including rather new systems consisting of low molecular weight
organic components, e.g. conducting organic solids and liquid
crystalline inclusions.

The properties of blends were also considered in terms of
models proposed for elucidation and prediction of the performance
of blends.

We are aware that the above topics do not cover all problems
which are of huge interest in the field of rapidly developing
science and technology of polymer blends. It seems however, that
the papers included in this volume do indicate some new trends
and perspectives for further research in the properties, problems
and opportunities of this class of materials.

The editors wish to express their gratitude to the authors
who have contributed to this volume and to the Polish Academy
of Sciences and Italian Research Council for their financial
support to the organization of the Second Polish-Italian Joint
Seminar on Multicomponent Polymeric Systems.

The editors are grateful to Mrs. Krystyna Krawczyk for her
secretarial help.

Finally, they appreciate the assistance and cooperation of
many people of the Centre of Molecular and Macromolecular Studies
of the Polish Academy of Sciences.

Centre of Molecular and   Marian KRYSZEWSKI
Macromolecular Studies
Polish Academy of Sciences   Andrzej GAŁESKI
Łódź, Poland

Istituto di Ricerche su   Ezio MARTUSCELLI
Tecnologia dei Polimeri
e Reologia, C.N.R.,
Arco Felice, Napoli, Italy

CONTENTS

RECENT ADVANCES IN INTERPENETRATING POLYMER NETWORKS

L. H. Sperling

Lehigh University
Materials Research Center
32, Bethlehem, PA

Recent research on IPN's has emphasized thermoplastic IPN's based on physical cross-links, and the factors controlling the variation of domain sizes in sequential IPN's. Most recently, decrosslinking and extraction studies on sequential IPN's has led to an improved understanding of the dual phase continuity sometimes present in these materials. The sequential IPN system poly(n-butyl acrylate)/polystyrene is emphasized.

INTRODUCTION

An interpenetrating polymer network, IPN, is defined as a combination of two polymers in network form, at least one of which is synthesized and/or crosslinked in the immediate presence of the other. An IPN can be distinguished from simple polymer blends, grafts, and blocks in two ways: (1) an IPN swells, but does not dissolve in solvents, and (2) creep and flow are suppressed[1].

Four basic types of IPN's can be distinguished: sequential IPN's, simultaneous interpenetrating networks, SIN's, thermoplastic IPN's and gradient IPN's. In forming a sequential IPN, the synthetic steps are taken in the following order: (a) polymer I is synthesized, (b) polymer I is crosslinekd, (c) monomer II plus crosslinker is swollen in, (d) monomer II is polymerized with crosslinking, and (e) phase separation between networks I and II takes place. In the above, (a) and (b) may be simultaneous or sequential in time, and usually are not distinguished. Step (e) is usually simultaneous

with (d), but starts after (d) has proceeded to the point where the
free energy of mixing becomes positive.

For SIN's, monomer II or (prepolymer II) is added before step
(b). Thus to a greater or lesser extent, the two networks are formed
simultaneously. Network I chains are stretched and diluted by network
II in a sequential IPN, but only diluted in an SIN, altering many
morphological and physical properties. Of course it is required
that the two polymerizations be non-interfering reactions, such as
by stepwise and chain kinetics.

In the case of thermoplastic IPN's, the crosslinks are of
a physical, rather than a chemical, covalent nature. Important
types of physical crosslinks include the hard blocks of a multiblock
copolymer, ionomeric sites, or crystalline regions in semicrystalline
polymers. Frequently, the polymers exhibit some degree of dual phase
continuity. In all such cases, the thermoplastic IPN's behave as
thermosets at use temperature, but as thermoplastics at some more
elevated temperature.

With the gradient IPN's[2-4], the composition is varied within
the sample at the macroscopic level. This is conveniently carried
out by soaking a sheet of network I in monomer II for a limited
period of time, and then polymerizing II rapidly, before diffusion
equilibrium can occur.

The field of IPN's and related materials has been reviewed
recently[1,5,6]. The purpose of this paper is to present recent
results from the Lehigh University Polymer Laboratory.

Thermoplastic IPN's

The patent interest in this field[7-11] reflects the potential
that industry feels for this field. By the use of physical bonds,
some degree of flow potential is maintained in the materials,
although some compositions[11] boecome truly thermoset in a last
curing reaction.

Dual phase continuity or phase inversion is controlled by two
factors[1,12]: the volume fraction of each component, and its melt
viscosity. Larger volume fractions or lower melt viscosities tend
to make that phase continuous. Obviously, equal volume fractions
and equal melt viscosities promote dual phase continuity.

In the studies at Lehigh[13,14] the thermoplastic elastomer
Kraton G served as polymer I (see Fig. 1). Styrene and methacrylic
acid were dissolved into the Kraton G, and polymerized in situ.
Upon neutralization with shearing, the poly(styrene-comethacrylic
acid) melt viscosity increased, and a phase inversion took place.

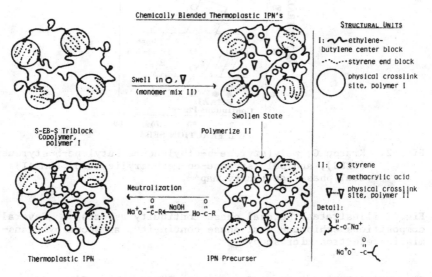

Fig. 1.   A chemically blended thermoplastic IPN from an SEBS triblock
copolymer and neutralized styrene-methacrylic acid copolymer.

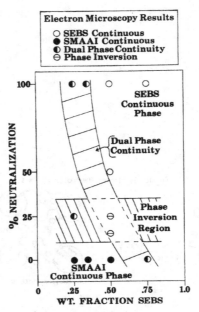

Fig. 2.   Kraton G, poly(styrene-b-ethylene-co-butylene-b-styrene)
          (SEBS), and poly(styrene-co-methacrylic acid) (SMAAI)
          dual phase continuity map[13].

Fig. 2 illustrates the dual phase continuity aspects[13]. Several
compositions exhibited dual phase continuity, as shown by transmission electron microscopy.

Sequential IPN's: Factors Determining Phase Domain Size

Because of the very small entropy of mixing and positive heat
of mixing, phase separation is the usual case in two component polymer mixtures rather than the exception. Far from being undesired,
phase separation frequently yields unexpected synergisms such as
toughening. Recent research in polymer blends, grafts, blocks,
and IPN's, suggests that the size of the domains, among several
other variables, is important in determining physical and mechanical
behavior. Experimentally, phase domain sizes range from a few
hundred angstroms to several microns, with polymer blends having
the largest domains and block copolymers and IPN's the smallest.

It is known that the domain size in block copolymers is
principally controlled by the individual block molecular weights,

and the ratio of their molecular weights. In the case of IPN's, the phase domain size of polymer II will be shown to depend on the crosslink density of polymer I, the volume fraction of each polymer, the interfacial tension, and the temperature.

Today, one of the most challenging areas concerning phase separated multipolymer materials relates to the prediction of the domain sizes. Various preparation techniques yield very different dimensions[1,15,16]. Much theoretical work on domain size prediction has been done on block polymer[17-20], as mentioned above, but not other multipolymer systems.

For IPN's, Donatelli et al.[21] earlier derived an equation especially for semi-IPN s of the first kind (polymer I crosslinked, polymer II linear) and extended this to full IPN's by assuming that the molecular weight of polymer II is infinite. Michel et al.[22] solved the Donatelli equation considering several boundary cases, and reinterpreted the constants involved. However, because of the semi-empirical nature of the Donatelli equation, its intrinsic shortcomings limit its applicability.

A new set of theoretical equations was proposed[23] which considered IPN's and semi-IPN's and also, under certain conditions, chemically induced polymer blends. The new equations contain no arbitrary constant, i.e. every term is experimentally measurable.

PROCESS PATH

Several assumptions are made for the derivation: (1) Thermo-dynamic equilibrium processes exist throughout the development of the domain formation. (2) The domains have identical diameters with a spherical shape. (3) The polymer networks obey Gaussian statistics. (4) A sharp interfacial boundary exists

The process path of domain formation is illustrated in Fig. 3. Initially, in state 1, network I is completely separated from monomer II (plus crosslinker). In state 2, polymer network I is swollen with the monomer II mixture. The path from state 1 to state 2 is accompanied by the mixing (dilution) of polymer I by monomer II, and concomitant expansion of polymer I chains caused by swelling with the monomer II mixture.

In state 3, polymers I and II are mixed and mutually diluted. Network I is stretched in the Flory-Rehner mode[24], although maximum swelling (with excess monomer) is not assumed. Demixing (phase separation) between polymer I and polymer II, with concomitant deformation of polymer II with further deformation of polymer I into a shell leads to state 4. State 4 shows a phase separated state, with a spherical domain of polymer II forming as the core,

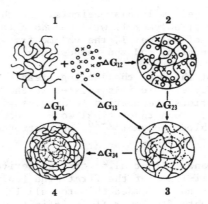

Fig. 3. A model of sequential IPN domain formations, showing the free energy changes involved[23].

surrounded by polymer I, deformed into a spherical shell. It should be noted that state 3 is imaginary. Phase separation begins during the polymerization itself. Obviously, the chains of polymer II do not move through the chains of polymer II. Fig. 3 shows that the average end-to-end distance of the polymer I chains increase in going from state 1 to state 4[21]. Very unexpectedly, this model also predicts that the polymer II end-to-end distance decreases during the same process.

The general equation[23] can be written in terms of the network II phase diameter, D_2:

$$D_2 = \frac{4\gamma}{RT(A\nu_1 + B\nu_2 - C)} \tag{1}$$

where

$$A = \frac{1}{2}\left(\frac{1}{\phi_2}\right)\left(3\phi_1^{1/3} - 3\phi_1^{4/3} - \phi_1\ln\phi_1\right) \tag{2}$$

$$B = \frac{1}{2}\left(\ln\phi_2 - 3\phi_2^{2/3} + 3\right) \tag{3}$$

$$C = \frac{\phi_1}{\phi_2}\frac{\rho_1}{M_1}\ln\phi_1 + \frac{\rho_2}{M_2}\ln\phi_2 \tag{4}$$

and ϕ_1 and ϕ_2 and ν_1 and ν_2 are the volume fractions and crosslink

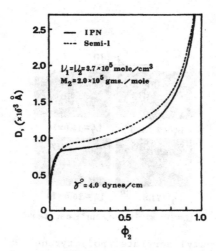

Fig. 4. Effect of composition ratio on the domain size of an IPN and a semi-1 IPN[23].

densities of polymers I and II, respectively and ρ_1 and ρ_2 and M_1 and M_2 are the densities and molecular weights of polymers I and II, respectively, provided they are not crosslinked. Thus terms A and B are the coefficients for ν_1 and ν_2, if not zero, and the first or second term (or both) in eq. (4) is used if either ν_1 or ν_2 or both are zero, respectively. As usual, RT represents the gas constant times the temperature.

It should be emphasized that equation (1) assumes the formation of spherical domains. There is a growing body of evidence[1,28], however, that some systems exhibit dual phase continuity. See below.

In the case of block copolymers, the calculated differences in domain dimensions between spheres and cylinders is usually less than about 50%. A similar difference is expected to materialize in the case of IPN's, when the proper analysis is completed.

The variation of D_2 and ϕ_2 is illustrated in Fig. 4 for a sequential IPN and a semi-1 IPN (only polymer I crosslinked). Typical molecular parameters were assumed. In the range of $0.1 < \phi_2 < 0.7$, it is interesting to note that the change in D_2 with ϕ_2 is modest. Domain diameters in the range of 700 to 1000 Å are predicted. Such values are commonly seen, experimentally[1]. Equation (1) also predicts a decrease in domain size with increasing crosslink density. While both networks exert important forces,

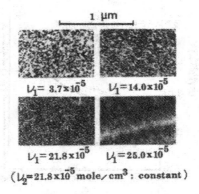

Fig. 5. Poly(n-butyl acrylate)/polystyrene IPN's. Phase domain size as a function of network I crosslink level[15].

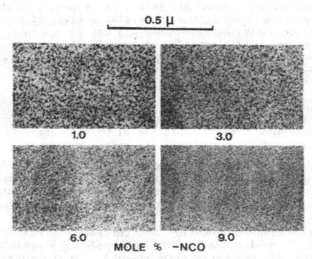

Fig. 6. Poly(n-butyl acrylate)/polystyrene IPN's. Phase domain size as a function of polymer I/polymer II graft level[26].

Table 1. Leading Factor Controlling Morphological Characteristics
(Besides interfacial tension)

System	Domain size	Miscibility
Block Copolymers	Molecular weight of block in domain	Absolute length of individual blocks
Sequential IPN's	Crosslink density of Polymer I	Graft level

a change in network I is about 6-10 times as important as an equivalent change in network II.

Fig. 5 shows how the domain size decreases with increasing crosslink density in network I for the system poly(n-butyl acrylate)/polystyrene[15]. A variation of Kanig's technique was used to stain the acrylic phase[25].

If the two polymers in an IPN are deliberately together, one would also expect changes in miscibility and domain size[26,27]. While there is good agreement about increasing miscibility with increased graft level, the two studies to date[26,27] have shown different directions for the domain size growth. The cause is not yet clear, but one system is an SIN, while the other[26] is a sequential IPN. The change in D_2 with increasing graft level for the sequential IPN is shown in Fig. 6[26].

Dual Phase Continuity in Sequential IPN's

One approach to the study of dual phase continuity in sequential IPN's is through the decrosslinking and subsequent dissolving out of one polymer or the other[28]. For example, in a recent study of a polypropylene/EPM morphology[29], the ethylene-propylene rubber component was dissolved out, leaving a sponge-like structure of polypropylene, as observed via scanning electron microscopy.

In a study of the sequential IPN system PnBA/PS, it was decided to crosslink the PnBA component with acrylic acid anhydride, AAA. On soaking overnight in warm ammonia water, the AAA hydrolyzes, decrosslinking the polymer. Then, Soxhlet extraction with toluene or acetone quantitatively removes the PnBA, leaving the crosslinked PS behind. Scanning electron microscopy was performed on the

Table 2. Extractable Material for Homopolymer Networks

Polymer	Crosslinker[a]	Sol fraction $(\%)$[b]
PnBA[c]	DVB	8
PnBA[c]	AAA	13
PS	DVB	6
PS	AAA	14

a) 1 mole - %
b) in toluene
c) 0.3 mole % 1-dodecanethiol added

remaining monolithic structures, as well as other characterization
studies[28].

The amount of extractable polymer, i.e. the percent of polymer
not incorporated into the network, is shown in Table 2. Employing
toluene, a good solvent for both polystyrene and poly(n-butyl acrylate), the sol fraction was relatively high, ranging from 6 to 14%.
This may be due to the low crosslink density of the samples. However for both polymer networks, higher amounts of extractables
were obtained when AAA is used as crosslinker.

In Table 3, experimental values of M_c were estimated from
swelling measurements and shear modulus at 10 seconds. The data
confirm the difference between divinylbenzene and acrylic acid
anhydride. When DVB is used as the crosslinker, the experimental
values agree well with the theoretical ones. The close agreement
may be fortuitous however, since one would expect physical crosslinks to contribute to the experimentally measured crosslink level,
especially when modulus was used. Apparently, network defects and
physical crosslinks nearly cancel each other out in this system.

For AAA on the other hand, the M_c values are much higher than
expected, indicating lower crosslink formation. Partial hydrolysis
of AAA could not explain the results since less than 5% free acrylic
acid was found in the AAA by infra-red analysis. An alternative
explanation for the lower than expected crosslink densities for AAA
crosslinked networks may be the possibility of ring formation.
It is well known that the corsslinking in diacrylate or diamethacry-
late systems is not complete, due to competition between intermolecular crosslinking and intra-molecular cyclization. In AAA, the

Table 3. Homopolymer Network Characterization Data

Polymer network	Average molecular weight between crosslinks, M_c						
	theoretical	from modulus data	from swelling data, solvent:				
			toluene	THF	benzene	MEK	acetone
PnBA (DVB)	13.000	14.400[a]	12.200	12.700	12.600	13.200	12.800
PnBA (AAA)	13.000	38.100[a]	36.200	35.900	37.400	41.300	42.600
PS (DVB)	10.500	10.800[b]	8.300	–	–	11.000	–
PS (AAA)	10.500	27.700[b]	21.900	–	–	25.800	–

a) at room temperature
b) at 160° C

distance between the two vinyl end-groups favors intramolecular cyclization.

After the decrosslinking step, the soluble polymer thus obtained was found to have a similar molecular weight to that of the corresponding linear polymer, prepared in the same experimental conditions, but without addition of crosslinker, around 3.5×10^5 g/mole for polystyrene, value which is in accordance with previous results from this laboratory. The difference in molecular weight between linear and decrosslinked poly(n-butyl acrylate) is around 10% when 0.3% chain transfer agent was added. The GPC chromatograms show a minor broadening of the molecular weight distribution for the decrosslinked polymer. This indicates that the major point of attack was a chemical degradation of the three-dimensional junction sites.

Sequential PnBA/PS IPN's

Sequential PnBA/PS IPN's were prepared with various compositions. Thus, the material was rigid where polystyrene was the major component, and soft when poly(n-butyl acrylate) dominated, confirming previous studies by Yeo et al.[15,26].

In the range 30/70 to 70/30, all the samples were opaque whereas for 80/20 and 90/10 PnBA/PS ratios, they were more transparent but still opalescent. This indicates a phase separated, heterogeneous system, as is well-known from earlier studies.

PnBA (AAA)/PS (DVB)

When soaked in ammonium hydroxide, network I is decrosslinked, and can be extracted (see Table 4). The amount of PnBA extracted roughly equals the amount in the IPN. By using toluene, the amount of extractable material is higher than expected, especially for high polystyrene content.

Some PS (from imperfect crosslinking of the polymer network II) is also extracted. For example, analysis by U.V. spectroscopy at 262 nm reveals 11.0% of PS in the sol fraction of a 50/50 decrosslinked IPN, which explains the high values obtained when toluene was employed as the extracting fluid.

Using acetone, which is a good solvent for the PnBA phase but a bad one for the PS, the extraction data in Table 4 show that at each composition the PnBA phase was quantitatively removed from the IPN. This means that a) the decrosslinking reaction occurs even when the PnBA network is in an IPN form (effective diffusion of the ammonium hydroxide in the network); b) the grafting level between

Table 4.   Results of Soxhlet Extraction for Decrosslinked
           PnBA (AAA)/PS (DVB) IPN's with Various Composi-
           tions

IPN weight composition PnBA/PS	Sol fraction %	
	toluene as solvent	acetone as solvent
0/100	6.1	–
30/70	37.8	29.5
40/60	46.4	38.3
45/55	48.8	43.8
50/50	53.2	49.6
55/45	62.7	56.1
60/40	69.2	61.6
70/30	75.9	71.2
80/20	85.1	83.0
90/10	94.4	94.6
100/0	100	100

polymer I and polymer II is low, probably less than five percent;
c) for all compositions, the PnBA phase must exhibit some degree
of continuity to be quantitatively removed. This last seems obvious
when polymer I is the major component but is not evident when polymer II predominates. That network I retains its phase continuity
under all conditions examined further indicates that network I
chains were not degraded during swelling with monomer II.

The optical properties and the macroscopic aspects of the
samples in the dry state after extraction are of certain importance:
surprisingly, for every composition the decrosslinked IPN was almost
transparent. This must be caused by the very small size of the
holes (see below). Note also that when the IPN had more than 30% PS,
a monolithic sample remained in the thimble; at 30%, after drying,
the material was very brittle. For the lower compositions studied
(10 and 20% PS) small grains were obtained. This means that for
midrange compositions, polymer II was also continuous. Roughly
speaking, the limit of polymer II continuity can be situated
between the 70/30 and 80/20 PnBA/PS ratios.

Table 5. Density measurements and Swelling Behavior for PnBA
          (AAA)/PS (DVB) IPN's and Decrosslinked IPN's with
          Various Compositions

IPN weight composition PnBA/PS	IPN density (g/cm^3)	Decrosslinked and extracted IPN		
		ρ (g/cm^3)	q_v a)	M_c b) $(g/mole)$
0/100	–	–	4.70	8300
30/70	1.053	0.984	5.78	13800
40/60	1.022	0.781	6.43	17800
45/55	2.079	0.845	6.33	17200
50/50	1.067	0.795	7.04	22000
55/45	1.059	0.818	10.37	52000
60/40	1.052	0.744	15.35	117000
70/30	1.023	0.965	18.96	178000

a) True swelling plus interstitial solvent held by capillary
   forces.
b) Apparent value estimated from apparent swelling.

    The remaining polystyrene network was characterized by swelling
studies. For each composition, when possible, the equilibrium
swelling degree in toluene and the apparent M_c values were calculated,
and are listed in Table 5. The M_c data are even higher than the
corresponding M_c value reported in Table 3 for homopolystyrene
crosslinked with DVB. In fact, one can not consider the extracted
and decrosslinked IPN, i.e. the remaining polymer II as a conventional PS network. The decrease in density (see Table 5) suggest a
porous material with some liquid imbibed in the pores during
swelling.

    In general, all scanning electron micrographs of decrosslinked
and extracted IPN's showed a complex structure and revealed the
internal appearance of a sponge with submicroscopic porosity. The
decrease in density (see Table 5), although less than theoretical,
confirms that structure. Fig. 7 and 8 show micrographs for a mid-
-range composition: in Fig. 7, the white portions represent the
remaining phase, which is polystyrene and the dark zones are voids
where network I was located. Undoubtedly, the PS phase is continuous
in space. The voids seem to be continuous too, and hence the
poly(n-butyl acrylate) phase must have been continuous before

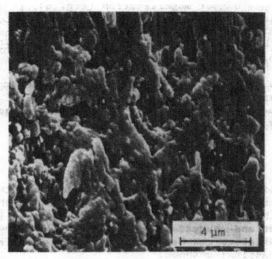

Fig. 7. Scanning electron micrograph of decrosslinked and extracted
A-type IPN [PnBA (AAA)/PS (DVB), 50/50], low magnification.

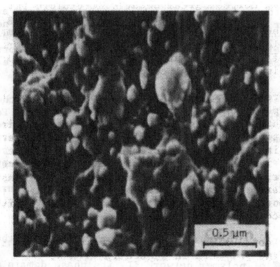

Fig. 8. Scanning electron micrograph of decrosslinked and extracted
A-type IPN [PnBA (AAA)/PS (DVB), 50/50], high magnification.

extraction. At a higher magnification (Fig. 8), it can be seen
that the continuous phase is formed by an agglomerate of spheres.
The diameter of these spheres is approximately 100 nm. This is
in accordance with the previous theoretical calculations and
transmission electron microscopy measurements.

For IPN's with only 30% PS, the decrosslinked material collapsed
during the drying step after extraction of polymer I and reveals
a cracked structure with grooves. This was suggested by the value
of the density found for that sample, see Table 5.

CONCLUSIONS

The use of labile crosslinks allows a different approach to
the study of interpenetrating polymer networks. With acrylic acid
anhydride as the crosslinker, hydrolysis leads to a linear polymer
easy to extract and characterize, and to a pure homopolymer network
whose characteristics can be compared with an identical network
prepared by classical methods.

First in the sequential IPN's, AAA was used to crosslink PnBA,
i.e. polymer I, and DVB to crosslink PS i.e., polymer II. Nearly
the whole amount of decrosslinked polymer can be extracted, meaning
a low level of chemical grafting between polymer I and polymer II.

Interesting morphological details heretofore not available
were found for the PnBA (AAA)/PS IPN's: the remaining network II
presents a porous structure as indicated by density measurements
and scanning electron microscopy. The voids, which correspond to
the location of the poly(n-butyl acrylate) phase before extraction,
are connected in the midrange compositions.

Most interestingly, the extent of continuity of the remaining
polystyrene depends on its concentration in the IPN. For mid-range
compositions, high magnification scanning electron micrographs showed
dual phase continuity on a supermolecular level. When the amount
of polymer II in the IPN was decreased below 30%, the sample crum-
bled by itself indicating a macroscopically discontinuous second
phase. However, at a microscopic level polymer II was still continuous to a large extent. It is when polymer II is around 10%
that the discontinuity is evident and only a few individual
particles are connected.

The major conclusion from this paper relates to the dual phase
continuity of the polymers in sequential IPN's:
i)   above 20% of polymer network II, its phase domain structure
      was continuous;
ii)   throughout the composition range studied, polymer network I
      was continuous.

REFERENCES

1. L. H. Sperling, "Interpenetrating Polymer Networks and Related Materials", Plenum Press, New York (1981).
2. G. Akovali, K. Biliyar and M. Shen, J.Appl.Polym.Sci., 20:2419 (1976).
3. C. F. Jasso, S.D. Hong and M. Shen, Polymer Preprints, 19(1):63 (1978).
4. G. C. Martin, E. Enssani and M. Shen, J.Appl.Polym.Sci., 26:1465 (1981).
5. Yu.S. Lipatov and L.M. Sergeeva, "Interpenetrating Polymeric Networks" Naukova Dumka, Kiev (1979).
6. D. Klempner, Angew.Chem., 90:104 (1978).
7. S. Davison and W.P. Gergen, U.S. 4,041,103 (1977).
8. W. P. Gergen and S. Davison, U.S. 4,101,605 (1978).
9. W. K. Fischer, U.S. 3,806,558 (1974).
10. F. G. Hutchinson, Br. 1,239,701 (1971).
11. F. G. Hutchinson, R.G.C. Henbest and M.K. Leggett, U.S. 4,062,826 (1977).
12. G. N. Avgeropoulos, F.C. Weissert, P.H. Biddison and G.G.A. Bohm, Rubber Chem.Tech., 49:93 (1976).
13. D. L. Siegfried, D.A. Thomas and L.H. Sperling, J.Appl.Polym.Sci., 26:177 (1981).
14. D. L. Siegfried, D.A. Thomas and L.H. Sperling, Polym.Eng.Sci., 21:21 (1981).
15. J. K. Yeo, L.H. Sperling and D.A. Thomas, Polymer Eng.Sci., 21:696 (1981).
16. D. Klempner and D.C. Frisch, eds., "Polymer Alloys II." Plenum, New York (1980).
17. D. J. Meier, J.Polym.Sci., Part C 26:81 (1969).
18. D. J. Meier, Polym.Prepr., Am.Chem.Soc., Div.Polym.Chem., 18:340 (1977).
19. E. Helfand, J.Chem.Phys., 62:999 (1975).
20. T. Inoue, T. Seon, T. Hashimoto and H. Kawai, Macromolecules, 3:87 (1970).
21. A. A. Donatelli, L.H. Sperling and D.A. Thomas, J.Appl.Polym. Sci., 21:1189 (1977).
22. J. Michel, S.C. Hargest and L.H. Sperling, J.Appl.Polym.Sci., 26:643 (1981).
23. J. K. Yeo, L.H. Sperling and D.A. Thomas, in preparation.
24. P. J. Flory, "Principles of Polymer Chemistry", Cornell, Ithaca (1953).
25. G. Kanig, Prog.Coll.Polym.Sci., 57:176 (1975).
26. J. K. Yeo, L.H. Sperling and D.A. Thomas, accepted J.Appl. Polym.Sci.
27. P. R. Scarito, L.H. Sperling, Polym.Eng.Sci., 19:297 (1979).

28. J. M. Widmaier and L.J. Sperling, submitted, _Macromolecules_.
29. E. N. Kresge, _in_ "Polymer Blends", Vol. 2, D.R. Paul and
       S. Newman, eds., Academic Press, New York, 1978.

ACKNOWLEDGEMENT

    The authors are pleased to acknowledge National Science
Foundation Grant No. MR-8015802, Polymers Program.

ADHESION BETWEEN POLYMERS

H.W. Kammer and J. Pigłowski*

Technische Universität Dresden, Sektion Chemie
DDR-8027 Dresden, Mommsenstrasse 13, GDR

*Technical University, Wrocław, Poland

INTRODUCTION

For most polymers it is thermodynamically unfavorable to form
homogeneous mixtures with each other. This is so because the combi-
natorial entropy of mixing of two polymers is dramatically smaller
than that for two low molecular weight compounds. The enthalpy of
mixing, on the other hand, is often a positive quantity or, at best,
zero. In such cases immiscibility results when polymers are mixed.
Many of the advantages of such multicomponent polymeric systems are
in fact direct results of this incompatible nature. There has recently been a great deal of interest in the studies of the structure
and properties of heterogeneous multicomponent polymers or polymeric
alloys[1]. In fact, immiscibility is desired in many cases since polymer-polymer composites in which each phase can contribute its own
characteristics to the product exhibit new properties. However, in
the solid state good mechanical behavior requires efficient transfer
of stress between the component phases, which depends on the adhesion
at the interface. More often than not, polymers do not adhere to each
other well, and poor mechanical properties of polymer mixtures
result. The consideration of adhesion between polymers ought to be
an integral part of developing of blend systems.

Before the adhesion between polymers can be discussed, it is
essential to clarify what "adhesion" means in physical terms. In
the simplest approximation, "adhesion" signifies the sticking
together of two different materials. From a molecular or microscopic
point of view adhesion is related to energetic quantities and is the
sum of all intermolecular interactions at the interface. This is
what Mittal[2] called "basic adhesion". In macroscopic terms, adhesion

19

means that force or work, in tension or shear, are transmissible
from one phase to the other, that is, mathematically, the stress
tensor has to be continuous across the interface. Therefore it is
necessary to characterize adhesion by the force or the work required
to effect separation of the adhering phases. Mittal[2] suggested the
term "practical adhesion" to represent the force or the work
required for breaking of adhesive joints. However, this is
not sufficient. All adhesion measurement techniques provide
quantities which are affected by the test rate. It is
evident that the main information on adhesion is obtained from
a destruction process or, generally speaking, from an irreversible
process. Irreversible processes are not only determined by forces
but also by rates. So, obviously, adhesion expressed in terms of
mechanical strength also depends on the rate of destruction. Only
in the limiting case of extremely low rates of failure are equilibrium conditions realized. As a direct consequence then, adhesion is
expressible in terms of thermodynamic or reversible work. Thermodynamic or reversible work of adhesion is the change in free energy
when the materials are brought in contact, and it is the same as
the amount of work expended under reversible or equilibrium conditions to disrupt the interface. We may conclude that there is no
simple relation between "basic adhesion" and "practical adhesion".
At the microscopic level, the existence of molecular interactions
is the condition for strength in an adhering system. "Basic adhesion"
depends exclusively on the interface characteristics of adhering
phases. "Practical adhesion", however, is affected by specimen
properties and by the measurement technique, that is the manner of
applying external forces and the rate of test. So, for a given
combination of materials, different techniques yield different
practical adhesion values. In conclusion, practical adhesive strength
is determined by the interfacial contact, the size of interfacial
defects, and the extent of irreversible deformations occurring in
the fracture region. It should be noted, however, that in practical
situations, one is concerned with "practical adhesion", not with
"basic adhesion".

Various theories to explain the mechanism of adhesion have been
proposed. Most of them deal with the formation of intimate interfacial contact. The problem is that adhesive strength depends not
only on the extent of intermolecular interactions at the interface
but also on the mechanical response of the materials. This lecture
will review some of these theoretical concepts and will close with
a discussion of our own investigations of adhesion between polymers.

THEORIES OF ADHESION

At the most sophisticated level the theory of adhesion ought
to explain the macroscopic properties of adhesive layers in terms
of molecular structure and behavior under mechanical stress

distributions. No such general theory exists. The wide range of
adhesion phenomena is divided into different aspects, and each
theory deals with only one aspect of the complex adhesion phenomenon.
Therefore, each of these theories is of limited validity. Theories
of adhesion are based on fracture mechanics, on surface properties
(adsorption and wetting), on diffusion, and on electrostatic or
chemical interactions.

The Fracture Theory of Adhesion

The starting point of the fracture theory is the Griffith-Irwin
theory[3] of cohesive fracture. It has been extended to the adhesive
fracture[4],[5]. According to this theory the fracture strength σ of an
adhesive bond is related to the fracture energy ε and the critical
crack length l. It is

$$\sigma \sim (E \, \varepsilon/l)^{1/2} \quad , \tag{1}$$

where E is the modulus of elasticity of the adhesive layer. The
fracture energy ε per unit area of the fracture surface is given
by the thermodynamic or reversible work of adhesion, W_A, and the
energy W_D dissipated in the irreversible processes during the crack
formation. We get

$$\varepsilon = W_A + W_D . \tag{2}$$

In the case of reversible separation, the term W_D will be negligible.
On the other hand, at high rates of separation the term W_D is very
much larger than W_A. According to Eq. (1), above a certain level of
loading some preexisting void starts to grow as a propagating crack.
We may thus conclude that the fracture process is determined by crack
formation and crack propagation. Griffith crack theory is essentially
a static conception of critical crack formation. Crack growth, however, also depends on dissipative processes. Below the critical load,
crack propagation may advance very slowly. In such a case there is
a dissipation of energy due to creep processes. Therefore, fracture
is a time-dependent process. This aspect is neglected in the
Griffith-Irwin theory of fracture.

It may be interesting to discuss the dissipation of energy
during an adhesive separation process. Mechanical deformation takes
place in the interfacial region or in regions near it. The thickness
of the layer in which the energy is dissipated can be roughly estimated by thermodynamic considerations. It is well known that the
dissipative processes in fracture release heat. The amount of heat
produced during the separation process depends on the rate of the
process. At sufficiently high rates the produced heat cannot be
conducted away, and the process is adiabatic. Thus, the energy
dissipated as heat must raise the temperature in the

fracture region. Following Good[6] we assume

$$\epsilon = \frac{1}{A} \int_{T}^{T+\Delta T} C dT = \frac{\bar{C}}{A} \Delta T, \qquad (3)$$

where $\bar{C}$ and A are the mean specific heat and the specific area of the deformation region, respectively. Using $A = (\zeta \tau)^{-1}$ we get for the thickness τ of the layer

$$\tau = \frac{\epsilon}{\bar{C} \zeta \Delta T} . \qquad (4)$$

ζ is the mean density of the fracture region. For the specific heat and the density we can use $C = 1$ J/g.K and $\zeta = 1$ g/cm^3, respectively. The energy required for adhesive separation of immiscible polymers is found to be of the order of 0.01 J/cm^2 [7] *. For the temperature rise it is reasonable to set $\Delta T = 200$ K. With these values we find: $\tau = 0.5 \times 10^{-4}$ cm. On the other hand, the thickness of the interfacial region between immiscible polymers is of the order of 1 nm[8]. We can conclude that the energy may be dissipated in regions much larger than the interfacial region. So, the local rise of temperature in the fracture region, especially near the crack tip, leads to a local increase in the plasticity of the polymer, as its chains become more mobile. This causes an increase in the rate of the crazing process which results in further enhancement of the rate of heat production, and so on until the crack propagation starts. This is an explanation of fracture in terms of a thermal feedback process. The role of the interface in the separation process is a subject of many controversies in the literature[9-14]. According to Bikerman[9,10], true interfacial failure practically never occurs. Fracture occurs in a weak boundary layer which may be near the interface. On the other hand, true interfacial failure has been experimentally observed[11]. The adhesive separation process depends on the level of loading, the rate of the deformation, and the structure of the interfacial region. It is obvious from the above discussion that the mechanism of the separation process is determined by dissipative processes in crack formation and crack extension. Crack propagation can take place only when sufficient energy is available to fill the requirements of the dissipative processes in craze formation. Hence, it is possible to prevent adhesive separation by stabilization of interfacial crazes.

In the interfacial region microcracks always exist. Depending on the level of loading these defects can be the starting point of

---

*The thermodynamic work of adhesion is about three orders of magnitude lower.

creep processes or crack propagation in the interface. As the above discussion shows, with increasing stress and at higher rates of deformation the energy is dissipated in a much larger region than the interfacial region. Then crack propagation can start to a larger extent also in the neighbourhood of the interfacial region. However, only in the case of strong interactions between the phases (chemisorption or polymer grafting) or in the case of notable interpenetration of the chain molecules will the interfacial region contribute to the enhancement of energy requirements for dissipative processes. This means that the strength of the transition region will be comparable with that of the bulk phases. Otherwise the strength of the interfacial region will be negligible, since interfacial crazing processes will not resist crack propagation as is the case with crazing in bulk.

The Wetting Theory of Adhesion

The specific thermodynamic work of adhesion, W_A, of two substances is equal to the sum of the two surface tensions less the interfacial tension

$$W_A = \gamma_1 + \gamma_2 - \gamma_{12} \tag{5}$$

The latter quantity is not directly measureable except in the case of liquids or melts. Essentially according to this theory, the energetics of wetting is responsible for the extent of interfacial contact and so for the adhesive strength. This is certainly a necessary but not necessarily a sufficient condition of good adhesion, since kinetic effects and much larger layers than the interfacial region may affect adhesion. The simplest condition of wetting liquid phase 1 on phase 2 is $\gamma_1 < \gamma_2$. Another useful equilibrium-based concept in connection with surface energetics is that outlined by Wu[14]. Wu suggested the spreading coefficient to be the driving force for wetting. This coefficient is defined as the thermodynamic work of adhesion less the work of cohesion of the spreading phase. Thus, the spreading coefficient λ_{12} for phase 1 on phase 2 is

$$\lambda_{12} = W_A - W_{C1} = \gamma_2 - \gamma_1 - \gamma_{12} \tag{6}$$

On the basis of the energy additivity concept it is assumed that surface and interfacial energies can be resolved into dispersion and polar components:

$$\gamma = \gamma^d + \gamma^p \tag{7}$$

Using Eq. (7) it has been shown by Wu that λ_{12} will have a maximum value when the relative polarities γ_i^p/γ_i of both components are equal. In such a case, adhesion should be favored. Wu has reported on qualitative correlations between λ_{12} and adhesive strength[14].

The importance of thermodynamically-based wetting theory of adhesion
is that it shows the possibility of forming an adhesive joint, though
it does not describe the load required to break a joint. It is
possible to combine the wetting and the fracture theory. The link
between them can be the interfacial void length 1. In accordance
with Eq. (6), for $\lambda_{12} > 0$ wetting should be rather perfect, and in
the opposite case it should be poor. Hence, length 1 of the unwetted
interfacial void can be related to λ_{12} in the following manner. Let
us consider the change in Gibbs free energy with respect to length 1
and the work of adhesion W_A, we get

$$dg = W_A \, d1 + 1 \, dW_A \quad . \tag{8}$$

If we impose condition γ_1 = const. on Eq. (8) and use Eq. (6), then
at equilibrium we obtain

$$\frac{d1}{1} = -\frac{d\lambda_{12}}{2\gamma_1 + \lambda_{12}} \quad . \tag{9}$$

Integration of this equation yields

$$1 = \frac{1_o}{1 + \dfrac{\lambda_{12}}{2\gamma_1}} \quad , \tag{10}$$

where 1_o is the void length at $\lambda_{12} = 0$. Substituting Eq. (10) in
Eq. (1) we get

$$\sigma \sim \left(\frac{E\varepsilon}{1_o}\right)^{1/2} \left[1 + \frac{\lambda_{12}}{2\gamma_1}\right]^{1/2} \quad . \tag{11}$$

This result differs slightly from that reported by Wu[15]. Expression
(11) takes into account the dependence of the adhesive strength both
on the mechanical properties of the system as E and ε and on the
wettability. Eq. (11) is probably a good approximation when the separation rates are low.

The Diffusion Theory of Adhesion

This theory is exclusively concerned with polymeric adhering
systems. The theory proposed by Voyutskii[16-18] establishes that
diffusion of polymer chains across the interface determines the
adhesive strength. It must be emphasized that diffusion of macromolecules across an interface is possible to a larger extent only
when polymer temperatures exceed their glass transition temperatures.

Furthermore, a fundamental prerequisite of this theoretical treatment
is that thermodynamic compatibility must exist between the components
at least in the interfacial region. As has already been mentioned,
most polymers are mutually incompatible. However, thermodynamic considerations show that segments of different chain molecules interpenetrate to various degrees to form an interfacial layer[19,20].
Let us review some elementary reflections with respect to compatibility and interfacial thickness. The necessary thermodynamic condition
for the miscibility of the different components is

$$\Delta F = \Delta U - T \Delta S < 0 \quad , \qquad (12)$$

where ΔF, ΔU, and ΔS are the free energy, the internal energy, and
the entropy of mixing, respectively. Let us consider two polymeric
components A and B with chain molecules each of r segmental units.
We take into account only the nearest-neighbour interactions between
the segments. In such a case, the energy of mixing is proportional
to the difference in the interaction energies between different (AB)
and equal segments (AA and BB). We can now introduce a dimensionless
parameter χ as a measure of this difference in the interaction
energies. Thus, a simple approximation to the energy of mixing is

$$\Delta U \sim kT \chi \qquad (13)$$

where k is Boltzmann constant. The entropy of mixing in the case of
low-molecular substances is simply proportional to k. For chain
molecules of length r, however, the entropy of mixing is diminished
by a factor of $1/r$, and we get

$$T \Delta S \sim \frac{kT}{r} \qquad (14)$$

Introducing (13) and (14) into Eq. (12) we obtain

$$\chi r \lesssim 1 \qquad (15)$$

as another criterion for mixing of two polymeric components.
If $\chi > 0$, this explains the fact that most polymer pairs are immiscible. In other words, the condition of miscibility is $\chi < 0$.
According to Eq. (15) polymers are incompatible for $\chi > 0$. However,
Eq. (15) provides for the possibility of an interfacial layer to
arise.   Interpenetration of the segments of chain molecules will
occur to such an extent as it is consistent with the expression (15).
We have to distinguish two cases.
1) In accordance with Eq. (15) it is possible, in the limit $\chi \to 0$,
that chain molecules interpenetrate in the interfacial layer. Then,
in terms of the Gaussian model, the extension of a chain molecule
with r segments, each of which has length b, is proportional to
$b \sqrt{r}$. Hence, the thickness τ of the diffuse interfacial region is
of the order of magnitude

$$\tau \sim b\sqrt{r} \quad (\sim \text{chain length}). \tag{16}$$

2) In the case when $\chi > 0$, chain molecules of type A will avoid contact with chain molecules of type B and vice versa. Thus, there is a loss of conformational entropy in the interfacial region because some conformations of A do not appear due to their high contact with B chains. In the interfacial region the chain molecules will have a deformed coiled shape. They are compressed. It is a very difficult problem to evaluate the entropy change arising from a distortion of polymer molecules from their equilibrium conformation. Thus, we simply introduce a compression coefficient α with $\alpha < 1$ into the Gaussian model. Then, the extension of a chain molecule in the interfacial region is proportional to $\alpha b\sqrt{r_s}$, where r_s is the number of segments entered in the interfacial layer. Using Eq. (15) we obtain for the thickness of the interfacial region

$$\tau \sim \frac{\alpha b}{\sqrt{\chi}} \quad (\sim \text{segment length}). \tag{17}$$

We may conclude that depending on χ there is possible a transition associated with the change from a diffuse interface to a sharp interface. Returning to the diffusion theory of adhesion we can establish that interdiffusion to a larger extent is only possible in case 1. Therefore, the adhesive strength should increase with decreasing value of the parameter χ. This has been observed by Voyutskii[21]. Parameter χ is proportional to the square of the difference in the solubility parameters δ. As shown in Fig. 1, the adhesive strength decreases with increasing disparity in the solubility parameters of the two polymers. So far we have only discussed the thermodynamic aspects of the diffusion theory. Therefore, the general statements of both the diffusion and the wetting theories have to be the same in this respect. With decreasing value of χ, the interfacial tension will decrease and, as is evident from Eqs. (6) and (11), λ_{12} and consequently the adhesive strength will increase. The diffusion theory of adhesion also involves kinetic aspects. Obviously, if there is diffusion of polymeric chains across the interface, then the adhesive strength will increase with increasing contact time. Thus, the time-dependence of adhesive strength can be expressed by a power law equation[22]

$$\sigma(t) \sim t^{\alpha}, \tag{18}$$

where α is a constant. This has been confirmed with α in the range of about $1/4$ to $1/2$[18]. However, it has been pointed out[23] that the kinetics of adhesive strength are adequately accounted for by rheological processes involved in producing intimate contact without diffusion being involved. Since the activation energies for the flow processes and for the diffusion are similar, wetting kinetics and diffusion exhibit nearly the same time-dependence. Therefore, it is difficult to separate both effects.

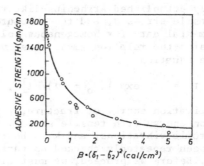

Fig. 1.   Relation between adhesive strength and solubility
          parameter difference. (Ref.[21]).

Theories of Electrostatic and Chemical Interactions
across the Interface

The electrostatic theory of adhesion[23] postulates that all
adhesion phenomena can be explained in terms of charge transfer
across the interface which results in the formation of an electrical
double layer. The double layer is comparable to a capacitor. Thus,
the work of adhesion is the work expended to separate the two layers
of this capacitor. In adhesion between polymers, electrostatic
effects probably play a less important role than other factors discussed above. Therefore, we will not go into details. If there is
a potential difference between the phases then we must add a further
term to the fracture energy ε of Eq. (2). It is well known that
interfacial chemical bonds can enhance the adhesive strength[25].
Experimental data show that several reactive functional groups such
as amino, hydroxyl, carboxyl, and also H-bond formation[26] tend to
increase adhesion. Adhesive strength is proportional to the amount
of functional groups. Thus, all that is needed for theoretical considerations is specification of the number of interfacial bonds and
their energy. The separation process is then considered as one
direction of the reaction of bond formation and breaking by the
application of stress. As we discussed above the problem is that
the probability of cohesive failure increases, when strong interfacial bonds exist. It is evident, however, that practical adhesion
can be improved by improving basic adhesion by interfacial chemical
bonds.

THE KINETIC CONCEPT OF ADHESIVE FAILURE

Now let us turn to our own studies on adhesion between polymers.
It has been pointed out by Zhurkov et al.[27-30] that fracture of
various materials is a kinetic phenomenon. This concept is based

on an experimentally established Arrhenius-like relation between the
lifetime τ, the tensile stress σ, and the absolute temperature T.
Analysis of experimental data for homogeneous solids indicates that
under uniaxial tension the relation among these variables is well
approximated by the equation

$$\tau = \tau_0 \exp \left[(U_0 - \gamma\sigma)/kT \right] , \tag{19}$$

where U_0 is the activation energy of fracture, γ symbolizes the
change of activation energy with tensile stress, $\tau_0^{-1} = 10^{13} \ s^{-1}$ is
the frequency of atomic vibrations, and k is Boltzmann constant.
Relation (19) has been repeatedly checked on various amorphous and
crystalline solids. Before we proceed, we must clarify the domain
of validity of Eq. (19). A more general formula, instead of Eq. (19),
is

$$\tau = \tau_0 \exp \left[U(\sigma)/kT \right] . \tag{20}$$

The generalized activation energy of fracture, $U(\sigma)$, may now be
expanded in a Taylor series and, since $\tau \to \infty$ at $\sigma \to 0$, we obtain

$$U(\sigma) = \frac{\alpha}{\sigma} + U_0 + (\frac{\partial U}{\partial \sigma})_{\sigma=0} \ \sigma + \ldots . \tag{21}$$

It follows from Eq. (21) that relation (19) is of limited validity.
Zhurkov equation becomes inapplicable in the limiting case $\sigma \to 0$ and
it is restricted to the region in which the activation energy $U(\sigma)$
is a linear function of the tensile stress σ. This approach, however,
permits a physical interpretation of fracture in terms of the activation energy U_0 and of the parameter γ. In the case of the fracture
of polymers, as has been reported by Zhurkov[28], U_0 is closely related
to the activation energy for thermal destruction of macromolecules,
and γ is proportional to the volume of a monomeric unit[31]. Clearly,
the question arises, whether Eq. (19) is appropriate to describe the
adhesive failure and, if it is, what would be the interpretation of
U_0 and γ. Thus, we wish to test the kinetic concept of fracture with
regard to adhesion between immiscible polymers. To measure the adhesive strength between polymers we used the pull-out method of Gorbatkina[32]. A bristle (about 1 mm in diameter) of one polymer was
embedded in a film of the second polymer cast from solution. The
specimens were stored in a dry box over a period of five days to
remove the solvent. After drying the thickness of the film was about
1 mm. The polymers used were branched poly(ethylene) (PE; density
0.913 g/cm^3 at 20° C), poly(ethyleneterephthalate) (PETP; density
1.336 g/cm^3 at 20° C), poly(styrene) (PS; density 1.045 g/cm^3 at
20° C), and poly(amide) (PA; density 1.136 g/cm^3 at 20° C). All the
listed polymers are commercial ones. The measurements included:
1) The lifetime of the adhesive joints under the influence of
different (constant) applied forces, and 2) the force required

Table 1. Activation energy U_0, parameter γ and critical tensile
stress σ_c of adhering polymer pairs and polymers

	U_0 kJ/mol	γ 10^{-6} m^3/mol	σ_c kPa
PE/PA	92	6870	930
PETP/PA	104	13610	1620
PE/PS	88	10290	950
from reference (31)			
PE	105	252	
PA	188	761	
PS	138	770 (33)	

to remove the bristle under conditions of a very small deformation
rate in a tensile tester (see values of σ_c in Table 1). The data
reported below are averages of about 20 measurements, with single
results being spread within ±30% of these values. The results of
lifetime measurements of five polymer-polymer systems are presented
in Fig. 2. Using Eq. (19) one has to expect a linear relationship
between lgτ and σ. The results for PETP/PA nad PE/PA are in accord
with Eq. (19). Deviations from Zhurkov equation are observed for the
system PE/PS at small tensile stresses. This is not surprising since
for $\sigma \to 0$ it has to follow $\tau \to \infty$, and Zhurkov relation is no longer
valid. At higher tensile stresses, there is a linear relationship
between lgτ and σ. For systems PA/PS and PETP/PS, we could not find
any relation among lifetime and tensile stress. In such cases it was
found that the lifetime is very high at small tensile stresses and
above a critical tensile stress it is very low. Activation energies
and parameters γ determined from the plots in Fig. 2 are presented
in Table 1. Furthermore, the corresponding values for fracture of
polymers are listed. It is obvious from Table 1 that the activation
energies of cohesive fracture of polymers are higher and the parameters γ are lower than the corresponding values of adhesive fracture. The polymer pairs investigated are incompatible. Hence the
interpenetration of chain molecules will be in the interfacial region
of that order of magnitude given by Eq. (17). In our opinion, an
adhesive joint between immiscible polymers breaks down due to molecular rearrangements and slippage of chain molecules in the interfacial region. Therefore, the activation energy U_0 is presumably
closely related to activation energies of molecular flow processes
taking place in the interfacial layer during the separation process.
The activation energies of viscous flow of poly(ethylene) and
poly(styrene), for instance, were observed to be about 54 kJ/mol[34]

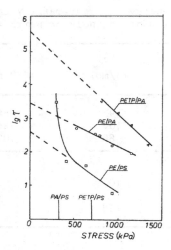

Fig. 2. Lifetime as a function of tensile stress for different
polymer-polymer systems at 20° C.

and 118 kJ/mol[35], respectively, at temperatures from about 150 to
250° C. If the mean value is employed in characterizing the flow
behaviour of the interfacial region we get 86 kJ/mol, and this
value coincides fairly well with U_o of Table 1. Owing to an abrupt
change in the activation energies of segmental motions with temperature, the lifetime increases with decreasing tensile stress as
shown in Fig. 2. In this respect, the immediate task in the development of the theory of adhesive strength is the establishment of
relationships between the activation energy U_o and the mechanisms
of cooperative molecular motions in the interfacial region. This
is an extremely difficult problem, so the foregoing considerations
must be regarded as somewhat conjectural.

        We will now turn to the interpretation of parameter γ. Zhur-
kov[28,31] simply related parameter γ to the volume V_m of a monomeric
unit:

$$\gamma = \beta V_m , \tag{22}$$

where β is the coefficient of overstress, which means that β signa-
lizes the distinction between the external load and the local stress
distribution. The lower the coefficient β, the smaller is the local
overstress, with the external load being distributed more uniformly.
Since Eq. (22) calls for a value of V_m; we chose the geometric mean
to calculate values of β for polymer pairs. The results are
listed in Table 2.

In our opinion it is interesting to interpret parameter γ in terms of the statistical fluctuation theory. Qualitatively, when the volume represented by γ is small, a high stress is necessary to start a fluctuation that is connected with the formation of a void and, therefore, the strength of the material is high. The average size of volume V_{fl} which ensures that a fluctuation associated with the crazing process will occur in the interfacial region can be calculated as follows. The starting point is the probability expression for fluctuations

$$p \sim \exp\ (-W_{min}/kT)\ , \tag{23}$$

where W_{min} is the minimal mechanical work necessary to pay for an entropy fluctuation that opposes the direction expected on a spontaneous pathway. The work W_{min} due to a change in the interfacial area can be expressed by changes in the free energy, ΔF, and the interfacial energy, so that

$$W_{min}\ =\ \Delta F(A) - \omega\Delta A\ , \tag{24}$$

where A and ω are the interfacial area and the interfacial tension, respectively. Since fluctuations are considered around the equilibrium value A_0 of the interfacial area, the term $\Delta F(A)$ can be expanded by a Taylor series around A_0 ($\Delta A \equiv A - A_0$). Thereby, terms higher than second order are neglected. Thus we get

$$W_{min}\ =\ \frac{\partial F}{\partial A}\ \Delta A + \frac{1}{2}\ \frac{\partial^2 F}{\partial A^2}\ (\Delta A)^2 - \omega\Delta A\ . \tag{25}$$

Obviously, the first and the third term cancel each other. Finally, it follows for W_{min}

$$W_{min}\ =\ \frac{1}{2}\ \frac{\partial \omega}{\partial A}\ (\Delta A)^2\ . \tag{26}$$

By analogy to the bulk phase, we define the modulus of elasticity of the interfacial area by

$$A\frac{\partial \omega}{\partial A}\ \equiv\ E_A\ . \tag{27}$$

This quantity can be related to the mean elastic modulus of the interfacial region as follows

$$E_A\ =\ E\ \tau\ , \tag{28}$$

where τ is the thickness of the interfacial layer. Using Eqs. (27) and (28), Eq. (26) can be rearranged to read

$$W_{min} = \frac{1}{2} \frac{E \tau}{A} (\Delta A)^2 . \qquad (29)$$

Thus, the mean square of the fluctuations of the interfacial area directly results from Eq. (23)

$$\overline{(\Delta A)^2} = \frac{k T A}{E \tau} . \qquad (30)$$

Now, we introduce local stress by

$$\sigma = E \frac{\Delta A}{A} , \qquad (31)$$

and use $V_{f1} = A \tau$ in Eq. (30). The result is

$$V_{f1} = \frac{k T E}{\sigma_c^2} , \qquad (32)$$

where σ_c represents the critical local stress. Thus, the quantity V_{f1} is that volume in which fluctuations inducing the crazing process will occur under the action of the local stress σ_c.
It should be noted, however, that Eq. (32) has been derived using Eq. (31), that is, only fluctuations associated with the elastic response are taken into consideration. In other words, Eq. (32) is restricted to situations close to equilibrium. This is approximately in accord with the definition (21) of γ, which means, that γ is the change in the activation energy barrier with stress in the limit $\sigma \to 0$. To calculate V_{f1} from Eq. (32) we will choose for the elastic modulus of the interfacial region the geometric mean, and replace σ_c by σ_c', the external critical stress. Thus, we get for the system PE/PA: $E = 3 \cdot 10^4$ N/cm^2, $\sigma_c' = 93$ N/cm^2 (from Table 1), and kT = $= 4.04 \cdot 10^{-19}$ Ncm. The calculation of V_{f1}' using the cited quantities gives a value of $1.4 \cdot 10^3$ nm^3, whereas the measured value of γ based on Eq. (19) is 11 nm^3 (see Table 1). In our opinion this large difference is due to the fact that the local stress appearing in Eq. (32) cannot be taken as equal to the external stress. Using the transformation $\sigma_c = \sqrt{\beta} \sigma_c'$, where σ_c and β represent the local stress and the coefficient of overstress, respectively, we obtain $V_{f1} = V_{f1}'/\beta = = 12$ nm^3 which is in good agreement with γ. Values of V_{f1} calculated in this way for other systems are listed in Table 2. The equality of parameter γ and of volume V_{f1} indicates that a connection exists between the decrease of the activation energy $U(\sigma)$ with increasing stress and the occurrence of fluctuations associated with the formation of voids in a certain volume.

In conclusion, the activation energy U_0 of adhesive fracture is the energy needed to induce cooperative molecular motions in

Table 2. The mean volume of monomeric units V_m, parameters β and γ, and the volume V_{fl} in accordance with Eq. (32)

Polymer pair	V_m (geom.mean) 10^{-6} m³/mol	β	γ nm³	V_{fl} nm³
PE/PA	58.5	117	11	12
PETP/PA	122.8	111	23	24
PE/PS	57.0	181	17	15

the interfacial region without external stress. When an outside force is applied, parameter γ determines the rate of decrease of the activation barrier with increasing stress. The statistical fluctuation theory can clarify the physical sense of the parameter γ. It is closely related to the very volume element in which crazing processes start due to fluctuations under external stress. Thus, the kinetic concept of fracture for homogeneous solids is conditionally valid for the adhesion between immiscible polymers. However, the molecular processes determing the quantities U_0 and γ are not really understood in detail. It remains a theoretical as well as experimental challenge to establish the relationships between the activation energy $U(\sigma)$ of adhesion and the molecular rearrangements in the interfacial region.

We have examined from different theoretical points of view the processes that occur in interfacial regions, in adhesion between polymers. An attempt has been made to link the observed macroscopic quantities, expressed in terms of U_0 and γ, and the molecular processes which establish the adhesive strength of polymeric systems.

REFERENCES

1. D. R. Paul and S. Newman eds., "Polymer blends", Vols. 1 and 2, Academic Press, New York (1978).
2. K. L. Mittal, Polym.Eng.Sci., 17:467 (1977); in "Adhesion Science and Technology", Vol. 9A, L.H. Lee ed., Plenum Press, New York (1975).
3. A. A. Griffith, Trans. Roy. Soc. London Phil., A 221:163 (1920); G.R. Irwin, in " Structural Mechanics", J.N. Goodier and N.J. Hoff eds., Pergamon Press, Oxford (1960).
4. R. J. Good, in "Recent Advances in Adhesion", L.H. Lee ed., Gordon & Breach, New York (1973).
5. M. L. Williams, in "Recent Advances in Adhesion", L.H. Lee ed., Gordon & Breach, New York (1973).

6. R. J. Good, in "Adhesion Science and Technology", Vol. 9A,
    L.H. Lee ed., Plenum Press, New York (1975).
7. J. Piglowski, H.W. Kammer, and C. Defer, Acta Polymerica,
    32:87 (1981).
8. H. W. Kammer, Z.phys.Chem. (Leipzig), 258:1149 (1977).
9. J. J. Bikerman, "The Science of Adhesive Joints", 2nd ed.,
    Academic Press, New York (1968).
10. J. J. Bikerman, J. Paint Techn., 43:98 (1971).
11. G. J. Crocker, Rubber Chem.Techn., 42:30 (1968).
12. R. J. Good, J. Adhesion, 4:133 (1972).
13. J. R. Huntsberger, J. Adhesion, 7:289 (1976).
14. S. Wu, J. Adhesion, 5:39 (1973).
15. S. Wu, in "Polymer Blends", Vol. 1, D.R. Paul and S. Newman
    eds., Academic Press, New York (1978).
16. S. S. Voyutskii, "Autohesion and Adhesion of High Polymers"
    (Russ.), Rostechisdat (1960).
17. S. S. Voyutskii and V.L. Vakula, J.Appl.Polym.Sci., 7:475 (1963).
18. S. S. Voyutskii, J. Adhesion, 3:69 (1971).
19. E. Helfand and Y. Tagami, J.Chem.Phys., 56:3592 (1972).
20. H. W. Kammer, Faserforsch. Textiltechn./Z. Polymerforsch.,
    29:459 (1978).
21. S. S. Voyutskii, S.M. Yagnyatinskaya, L.Y. Kaplunova and
    N.L. Garetovskaya, Rubber Age, 37 (1973).
22. R. M. Vasenin, Vysokomol.Soedin., 3:679 (1961).
23. J. N. Anand, J. Adhesion, 5:265 (1973).
24. B. V. Deryaguin and N.A. Krotova, Dokl.Akad.Nauk SSSR,
    61:849 (1948).
25. A. Ahagon and A.N. Gent, J.Polym.Sci.Phys., 13:1285 (1975).
26. W. H. Pritchard, in "Aspects of Adhesion", D.J. Alner ed.,
    Vol. 6, CRC Press, Cleveland Ohio (1969).
27. S. N. Zhurkov, Z.phys.Chem. (Leipzig), 213:183 (1960).
28. S. N. Zhurkov and V.E. Korsukov, J.Polym.Sci.Phys., 12:385 (1974).
29. V. R. Regel', Faserforsch. Textiltechn./Z. Polymerforsch.,
    29:8 (1978).
30. S. N. Zhurkov, K.J. Friedland, and V.I. Vettegren,
    Faserforsch. Textiltechn./Z. Polymerforsch., 26:126 (1975).
31. V. R. Regel', and A.I. Sluzker, in "Fisika savtra i sevodnya"
    (Russ.), Isd. Nauka, Leningrad (1973).
32. J. A. Gorbatkina, Vortragsmanuskripte der II. Internationalen
    Tagung uber glasfaserverstarkte Kunstoffe und Giessharze,
    DDR, Berlin (1967), (Germ.).
33. V. E. Til' and V.N. Kuleznev, "Struktura i mechanitseskie
    svoistva polimerov" (Russ.), Isd.vis.skol., Moscow (1972).
34. W. Philippoff and F.H. Gaskins, J.Polym.Sci., 21:205 (1956).
35. R. F. Boyer, Encl.Polym.Sci.Technol., 13:359 (1970).

PRINCIPLES AND TECHNOLOGIES OF POLYMER BLENDING

Sergio Danesi

Montepolimeri
Centro Ricerche Giulio Natta
44100 Ferrara

INTRODUCTION

A common feature of all commercially available blending equipment (extruders, mills, internal mixers) is that polymer components have to experience a shearing deformation process in the molten state. Under the action of the shearing force, the dispersed elements are elongated and broken down as schematically shown in Fig. 1.

Size reduction of the dispersed domains continues till the total force of the particle equals the surface tension which is the retracting force of the process.

A mathematical treatment of this process has been given by Taylor[1,2]. The continuous phase with a viscosity of η_o and applied shear rate of dv_x/dy deforms a sphere of radius R and viscosity η_1 with the force

$$T_\eta = C \frac{dv_x}{dy} \eta \ f(\eta_o/\eta_1)$$

where:

$f(\eta_o/\eta_1)$ – quantifies the effectiveness of the deformation process of the dispersed particle by the continuous phase and increases with the increase of the η_o/η_1 ratio.

The retracting force resulting from the surface tension is

$$T_\gamma = \frac{2\gamma}{R} \ .$$

Fig. 1. Schematic representation of particle disruption under
a shearing stress field.

The equilibrium radius of the dispersed particle occurs when
$T_\eta = T_\gamma$ and is equal to

$$R = \frac{c' \, \gamma}{\eta_o \frac{dv_x}{dy} \, f(\eta_o/\eta_1)}$$

From this equation it comes out that the equilibrium radius
is smaller when:
- the surface tension of the dispersed phase is lower
- the viscosity difference of the continuous and dispersed
  phase is higher
- the shearing forces are higher.

These relationships have been found particularly correct when
the amount of the dispersed phase is quite low and each particle is
isolated from the other. In this situation the competitive process
of particle coalescence, always present in unstable polymeric hetero-
phase systems, is drastically reduced indeed. Taking into account this
set of requirements, proper technologies and formulations can be
developed to control the morphology and characteristics of polymer
blends. This paper reviews the results of our research activity on
a typical heterophase systems based on isotactic polypropylene (PP)
and ethylene-propylene rubber (EPM) blends.

VISCOSITY DIFFERENCE OF COMPONENTS AND BLEND COMPOSITION

As shown in Fig. 2, commercially available polypropylene grades
have a melt viscosity which is either lower than, or equal to, that
of ethylene-propylene rubbers[3].

Two set of blends, having respectively completely different
and almost matching component melt viscosities, were prepared under
controlled and constant mixing conditions. The dispersion state of
the blend phases and its dependence on the blend compositon and melt
viscosity difference of the components is schematically summarized
in Fig. 3.

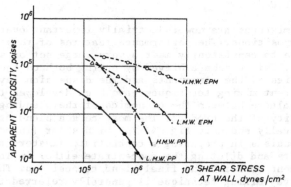

Fig. 2. Capillary rheological curves of base materials.
(Extr. temper. = 200° C).

Both series of blends exhibit a phase inversion for intermediate
compositions. The phase inversion region is narrow in the case of
blends with a high difference of the components melt viscosity, while
being very broad for matching viscosities. In this case one can also
observe the presence of continuous phases for a large composition
range. Blends with a greater disparity in the components melt viscosity show larger domains as in particular the rubber is the dispersed phase. As the amount of rubber in the blend increases, particularly beyond the phase inversion point, the viscosity difference
becomes less important in determining the dispersion state. All
blends with composition ratios of 80% rubber show a fine dispersed
phase structure which is essentially independent of the difference
of the components melt viscosity.

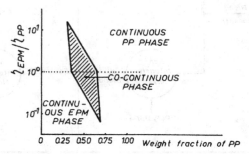

Fig. 3.   Schematic representation of morphology variation with
component viscosity ratio and blend composition.

MIXING TECHNIQUES

PP-EPM mixtures are now industrially important over the full
range of compositions. The performance features of this family of
products have to consistently meet a very large set of service
requirements. This can be achieved by acting both on the structural
characteristics of the components and the dispersion level of the
blend. Different mixing techniques have been developed to find out
the proper balance between the economics of the blending process
and the quality of the product. The most common blending techniques
are schematically shown in Fig. 4. The high shear (intensive) mixing
technique consists in preparing a concentrate (master-batch) of polymer components and diluting the concentrate either in low or high
shear equipment to reach the final blend composition. The low shear
(non intensive) mixing technique is generally referred to a direct
blending process of the final composition in a low shear equipment.

Particular operating procedures can be followed to promote
a higher degree of polymer dispersion. So, for instance, when the
mixer is held at the lowest practical temperature above the melting
point of PP, a better dispersion is normally achieved.

In a shear environment, elastomers, and in particular EPM elastomers, show only a moderate viscosity decrease as the temperature

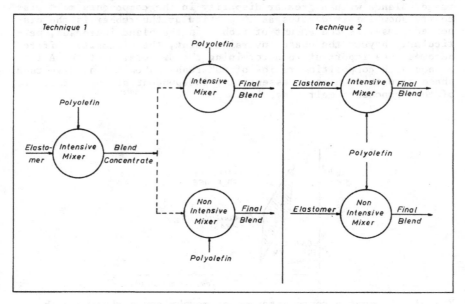

Fig. 4.   Blend preparation techniques.

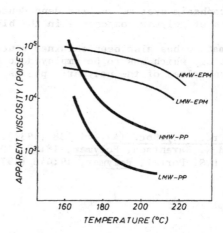

Fig. 5.   Viscosity-temperature relationship for selected PP
and EPM grades.

increases. PP, on the other hand, characteristically undergoes a rapid decrease of viscosity as the temperature increases (Fig. 5).
The crossover of the two curves occurrs normally just above the PP
melting point.

In this temperature range, the viscosities of the two components
are very similar and the mechanical shear can be effectively trans-
ferred to the mixture to enforce a good dispersion of the phases.
Intensive mixing and high homogeneization levels are especially
important when, for quality reasons, the modifying capability of the
elastomer has to be optimized. This is the case, for instance, of
the new family of elastomer modified polypropylene which has opened
a major market for plastic materials in the field of the automotive
bumpers. For this highly innovative utilization a severe set of properties has to be ensured by the material and in particular:
- considerable impact strength and ductile behavior in the entire
  range of car operative temperatures;
- good dimensional stability of the molded items at high temperature.
As these characteristics are differently influenced by the addition
of the rubber to PP, particular attention has to be paid when
choosing the right grade of elastomer and the proper degree of
dispersion in the plastomeric matrix.

CONCLUSION

PP-EPM systems have been shown to be particularly suitable to
experimentally confirm theoretical predictions about the dispersion

state of the heterophasic systems and its dependence on the structural characteristics of polymer components in the blend.

Useful information has also been obtained concerning the most appropriate technology which has to be employed to accomplish the prefixed dispersion values of the polymeric phases.

REFERENCES

1. G.J. Taylor, Proc. Roy.Soc. (A), 41:138 (1932).
2. D. Heikens and W. Barentsen, Polymer, 18:69 (1977).
3. S. Danesi and R.S. Porter, Polymer, 19:448 (1978).

THE ROLE OF INTERFACIAL AGENTS IN POLYMER BLENDS

G. Maglio and R. Palumbo

Instituto di Ricerche su Tecnologia dei Polimeri
e Reologia del C.N.R.
Via Toiano 6, 80072 Arco Felice, Napoli

INTRODUCTION

A considerable amount of research has been made over the last several years with a view to obtaining new polymeric materials with enhanced specific properties for specific applications or a better combination of different properties. After the synthesis of polymers from new monomers had been largely explored, efforts were focused on multiphase polymeric systems such as block or graft copolymers, composites, polymer blends, and interpenetrating networks [1-5]. Much attention is currently being devoted to the simplest route for combining outstanding properties of different existing polymers, that is, the formation of polymer blends [6,7]. Although an increasing number of miscible blends is reported in the literature [8], most polymer pairs are nonetheless immiscible thus leading to hetero-phase polymer blends [9,10].

The formation of two-phase systems is not necessarily an unfavourable event since many useful properties, characteristic of a single phase, may be preserved in the blend while other properties may be averaged according to the blend composition. Proper control of overall blend morphology and good adhesion between the phases are in any case required in order to achieve good mechanical proper-ties. It is well established that such parameters are strongly dependent on the presence in the blend of appropriate block or graft copolymers having chain segments chemically identical or similar to the homopolymers to be mixed. Such copolymers can be either preformed and added to the binary blend or formed "in situ" during the blending process. The influence of these copolymers, generally referred to as "interfacial agents" or "compatibilizers", has been related to their tendency to be preferentially located at

the interface between phases and to the capability of their individual chain segments to penetrate into the phase to which they are
chemically identical or similar [11,12].

Much work is currently being devoted to studying the influence
of parameters such as composition, copolymer architecture, and
segment size of interfacial agents on the morphology and on the
mechanical properties of two-phase blends.

MOLECULAR DESIGN OF INTERFACIAL AGENTS

As has been pointed out in a recent review by Paul and Barlow [8],
while some consequences of the use of interfacial agents in binary
blends are only qualitatively understood, it is quite clear that
their beneficial effects are strongly dependent on their molecular
design. When an AB block or graft copolymer is added to an A/B
binary blend, thus leading to a A/B/AB ternary blend (A and B are
two immiscible homopolymers), the following major effects have been
observed and studied, mainly in the field of toughened plastics:
i/ higher degree of dispersion; ii/ better adhesion between phases;
iii/ stabilization of microphases with regard to coalescence phenomena. Morphological observation of the three effects reported above
and property improvements in keeping with them can provide indirect
evidence of the influence of the addition of the third component,
i.e. the copolymer.

Heikens et al. [13,15] observed, in the case of 75/25 PS/PE
blends having 0%, 1%, 7% of added graft copolymer, that the dimensions of the PE domains decrease with increasing copolymer concentrations. Similar results were obtained for the same systems by
Paul [16], and for PS/Polydiene [17], for PVC/PE [18,19], and for Poly-
amide/Polyolefin [20-25] blends by several reserach groups.

Better adhesion between phases will result in improved
ultimate mechanical properties. Selected examples of this will be
given later. Direct morphological observation that block and graft
copolymers promote interfacial interactions between phases was made
by Heikens et al. [13] who reported the results of a fractographic
analysis of some PS/LDPE blends. In the fracture surface of binary
blends, the PE particles present smooth surfaces with no adhesion
to the PS continuous phase, while by adding 1% of graft copolymer,
surface irregularities are developed and may be attributed to PS/PE
connections broken during the fracture process. Similar evidence
of interphase adhesion was obtained in the case of Polyamide 6/EPM
blends having EPM-g-Polyamide 6 copolymer as the third component [25].

The third effect, i.e. the stabilization of fine dispersions
of polymer mixtures, may be of great importance when blends of
desidered morphology show a tendency to increase the particle

dimensions by coalescence phenomena during subsequent processing. The pioneering work by Molau [26-28] has clearly shown the emulsifying role of block and graft copolymers; in fact, their presence in mixed concentrated solutions of two immiscible polymers strongly delays phase separation. Several papers [29-30] have shown the emulsifying effect of copolymers in the case of PS/polydiene systems. There is thus quite considerable evidence that copolymers can act as interfacial agents.

Considering the question from a qualitative point of view, it is to be expected that the molecular architecture of the AB copolymer and the molecular weight of the individual segments will play an important role in determining the emulsifying efficiency of the copolymer. Fig. 1 presents a schematic picture of different copolymers located at the interface. It seems evident that, because of conformational requirements, the simpler the copolymer architecture, the easier the penetration of segments in the respective homopolymer phase. Consequently, block copolymers are expected to be more efficient than graft copolymers and diblock copolymers should be superior to triblock ones. Multiblock and multigraft copolymers should have the lowest interfacial effect. Experimental evidence in keeping with these considerations has been obtained by Riess et al. [29] in the case of PS-PI copolymers for DMF/hexane oil-in-oil

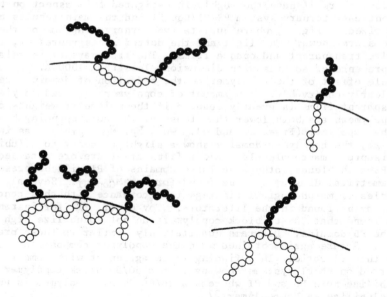

Figure 1.   Schematic picture of different types of copolymers acting as "interfacial agents".

emulsions and, recently, by Fayt et al. [31] in the case of PS/PE blends
obtained by melt mixing. As far as random copolymers are concerned,
no emulsifing behaviour is expected. Ethylene/propylene (EPM) statistical copolymers, however, have been found to improve the compatibility in LDPE/PP blends, with corresponding beneficial effects on
their mechanical properties [32]. Similar results have also been
reported for HDPE/PP/EPM blends [33]. This effect is more noticeable
with the use of ethylene rich EPM copolymers, the residual crystallinity of which may be ascribed to small polyethylene blocks. Also
in this case the interfacial role of the copolymer must be related
to a blocklike structure of the copolymer.

Other molecular parameters of AB copolymers important for their
emulsifying efficiency are the composition and the molecular weight
of individual segments. Since the best conditions are met when the
interfacial agent is concentrated at the phase boundaries and is able
to accept the homopolymer into its domains, the above parameters
should be such that the copolymer segregates in two phases, is not
soluble in either of the homopolymer phases, and is able to solubilize the homopolymers in its corresponding phases. It is now firmly
established that for a given AB copolymer the attainment of the
above conditions is directly dependent on the copolymer composition
and on the ratios of molecular weights of A and B segments of the
copolymer to those of A and B homopolymers respectively [34,35]. Riess
and co-workers [17] have thoroughly investigated this aspect on the
solvent cast ternary system PS/PI/Cop SI and the main results are
summarized in Fig. 2, where qualitative ternary diagrams of these
systems are shown. The light and the dotted zones represent, respectively, transparent and opaque films. The transparency is related to
the presence of domains with dimensions less than 1000 Å. A considerable effect of the copolymer on the reduction of domain sizes
is clearly observable. The amount of copolymer required to yield
transparent films is greatly reduced if the molecular weights of the
homopolymers are both lower than those of the corresponding blocks
in the copolymer (Figs. 2a and b). When $\overline{M}_{PI} > \overline{M}_{Cop\ SI} > \overline{M}_{PS}$, as in
Fig. 2c, the copolymer domains show a slight tendency to solubilize
polyisoprene macromolecule. Clear films are therefore obtained only
for PS rich blends, otherwise large domains of PI will be present.
A symmetrical diagram will be found for $\overline{M}_{PS} > \overline{M}_{Cop\ SI}$. Several
examples of morphological findings in accordance with this conclu-
sion may be found in the literature. Toy et al. [36], for instance,
have found that SBS triblock copolymers do not solubilize high M_w PB
in the PB domains. Diagrams qualitatively similar to those presented
in Fig. 2c are again obtained when the copolymer composition is far
from the 1:1 ratio. This finding is in agreement with some results
reported on PB-PI systems showing that a 50/50 block copolymer can
solubilize both PB and PI whereas a 34/66 PB/PI copolymer is unable
to solubilize PB homopolymer [37].

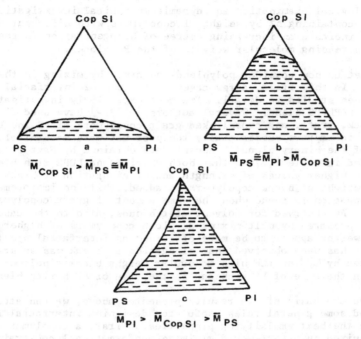

Figure 2.   Qualitative ternary diagrams of PS/PI/Cop SI blends.

Interesting information relevant to this point can be obtained from ternary blends of two different diblock copolymers with one corresponding homopolymer. Riess et al. [38] investigated the system PS/Cop SI$_1$/Cop SI$_2$ where Cop SI$_1$ and Cop SI$_2$ are diblock copolymers with M_w Cop SI$_1$ > M_w Cop SI$_2$. It was found that the Cop SI$_2$, of higher M_w, can act as an emulsifier for Cop SI$_1$ which forms the dispersed phase of the blend. Multilayered structures ("onion" structures) were easily obtained for these systems.

All these results show how molecular weight and composition influence the overall morphology of ternary blends having one of the components acting as interfacial agent.

Eastmond and co-workers [39] reached similar conclusions on quite different solvent cast blends consisting of PC, PMMA, and A$_2$BA$_2$ crosslinked block copolymers A=PC, B=PMMA of controlled architecture. The molecular weight of PMMA was in every case smaller than that of MMA blocks, whereas the molecular weight of PC blocks could be equal, larger, or much larger than that of PC homopolymer.

Morphological examination and dynamic mechanical investigations on
blends containing 5% by weight of copolymers with different M_w of PC
blocks indicate an increasing degree of homogenization in the blends
with increasing molecular weights of the PC blocks.

Let us now consider polyblends obtained by mixing in the molten
state. In this field, a few cases of the use of interfacial agents
have been studied in detail. The most extensively investigated system
is the LDPE/PS blend. Several authors [13-16,31] have used diblock,
triblock, graft, and crosslinked graft copolymers of PS and LDPE,
mainly to improve substantially the poor ultimate mechanical properties of the binary blends. Recent data obtained by Fayt et al.[31],
reported in Table 1, show that both PS rich and LDPE rich blends
exhibit higher values of strength and of elongation at break when
9% by weight of block copolymers is added, while no improvement
in elongation is found when the same amount of graft copolymer is
added. As observed for solvent cast blends, also in the case of
blends prepared by melt mixing do block copolymers of higher molecular weight appear to be more efficient as interfacial agents. In
fact it has been observed that elongation at break was substantially
affected by adding the high molecular weight block copolymer 2,
both in the case of PS rich and in the case of LDPE rich blends.

On the basis of the results presented above, we can attempt
to find some general rules useful for designing interfacial agents
having the best emulsifying properties. First, a copolymer structure
is required such that would minimize conformational constraints.

Table 1.   Improvement of ultimate mechanical properties of LDPE/PS
           blends by addition of block or graft copolymers
           (from ref. [31])

LDPE/PS composition	copolymer added (9% wt.)	ε_b improvement	ε_b improvement
70/30	graft	100	–
"	block 1	70	900
"	block 2	90	3000
30/70	graft	80	–
"	block 1	120	120
"	block 2	60	1300

$\overline{M}_n$: LDPE = 40 000, PS = 100 000, block 1 = 58 000,
    block 2 = 155 000.

Secondly, the M_w of individual segments have to be equal to or higher than those of corresponding homopolymers. With respect to this last point, it appears that the higher the molecular weight of the interfacial agent, the larger the amount necessary to saturate the interphase boundaries. The question that arises now is whether the above requirements will also help the copolymer to concentrate at the interface. The evidence presented by Riess et al.[40] seems to indicate that block copolymers show a marked tendency to be located preferentially at the interface when the ratios of the blocks M_w to the corresponding homopolymers M_w are nearly equal for the two components. When these ratios are both higher than one, i.e. copolymers of higher molecular weight, a certain amount of copolymer seems to be dispersed in the homopolymer phases. Heikens [13] investigated this point for the case of LDPE/PS blends by preparing several block and graft copolymers of different microstructure reported in Scheme 1. It was found that the graft copolymers contribute to the modulus according to their PS and PE content giving a single master curve in modulus-composition plots, independently of the ratio of bound to free PE, which is consistent with a simple two-phase system with the copolymer located at the interfaces. On the other hand, the addition of a block copolymer of regular structure and high M_w such as BC-1 results in modulus values lower than those of binary LDPE/PE blends of identical overall composition, and these values depend on the copolymer content. Bound PS does not contribute to the modulus in this case and it was therefore concluded that the block copolymer is not only located at the interface but that it forms a third, separate phase.

Scheme 1. Microstructure of PE/PS block and graft copolymers used in PE/PS blends (from ref. [13])

BC 1   PS ——— PE
              65 000   48 000

BC 2   PS ——— (PS-PE) ——— PE
              14 000   10 000   54 000

BC-3   PS ——— (PS-PE) ——— PE
              22 000   22 000   25 000

GC 1   PE   45 000
              ─────────────────
                             PS 10 000

GC 3   partly crosslinked graft copolymer

GC 4   gelled graft copolymer

Similar conclusions hold for the block copolymers of partially
disordered structure and lower M_w, BC-2 and BC-3, which show moreover a tendency to form less defined microphases and a continuous
soft phase at high concentrations.

Consequently, in the design of an interfacial agent, a balance
between conflicting requirements has to be found. In fact, the
conditions which should be met in order to optimize the emulsifying
properties of the copolymer may result in its lower concentration
at the interface. It is interesting to note that recent results
obtained by Eastmond and Phillips [41] suggest that multicomponent
polymers, including block copolymers, should in principle be immiscible with homopolymers of the individual components, unless the M_w
of the blocks are at least one order of magnitude larger than those
of the corresponding homopolymers. As these observations refer to
equilibrium conditions, they do not necessarily contrast with other
conflicting experimental evidence, but in any case they do suggest
that more research is needed in this field and that great attention
should be devoted to the kinetic aspects connected with blend preparation.

INTERFACIAL AGENTS IN TOUGHENED PLASTICS

Many industrial developments regarding polyblends, as well as
a great deal of fundamental research, concern the so-called "rubber
toughened plastics" [42]. In fact, most thermoplastic polymers widely
used especially in engineering applications show marked limitations
when toughness and high impact resistance are required. These characteristics have been generally improved by blending the thermoplastics with a small amount of a rubbery polymer. The resulting
material consists of a rigid matrix of high glass transition temperature in which a soft phase with a low T_g is dispersed. The finely
distributed soft particles will transform locally into heat the
mechanical energy supplied by the impact, by means of deformation
mechanism such as multiple crazes or shear bands initiated and
stopped at interphase boundaries. With respect to the parent thermoplastic, the blend will offer considerably improved ultimate mechanical properties and will suffer from a limited decay of modulus
and tensile strength. The overall result is a much better balance
of properties. As stated in the previous section, the presence of the
rubbery dispersion will however be effective in the toughening
process only if the rubber domains are of optimum size and adhere
strongly to the matrix. These conditions may be achieved with the
help of suitable block or graft copolymers.

The leading rubber modified plastics, such as HIPS and ABS, are
obtained by polymerization of styrene or of a styrene-acrylonitrile
mixture in the presence of polybutadiene. The outstanding properties
of these products result from the formation, during this process,

of a poly(butadiene-g-styrene) graft copolymer acting as an
oil-in-oil emulsifier of PS/PB or SAN/PB pairs.

Graft copolymers acting as interfacial agents may also be formed
during the melt mixing of the thermoplastic and rubber homopolymers
by means of mechanochemical reactions promoted by high shear forces
or by means of reactions such as interchange reactions in condensa-
tion polymers, or reactions of functional groups present as chain
ends or grafted species. In order to investigate the role of the
interfacial agent in rubber modified plastics – and, in particular,
to quantitatively investigate the influence of the microstructure,
of the copolymer on its interfacial activity – well defined block
and graft copolymers could be preformed and then added as separate
chemical species to the homopolymers during the blend processing.

The LDPE/PS system does offer these advantages and has been
thoroughly investigated by several research groups [13,15,31]. A clear
dependence of the impact strength of PS/PE/PS-g-PE blends on variations in the ratios of graft to free PE was found by Heikens et al.[13].
The addition of the graft copolymer increases significantly the impact
strength values of the binary PS/PE blend at any composition. Moreover, greater improvements in the impact strength were observed for
a fixed blend composition. Accordingly, the best results were found
for the binary PS/copolymer blend where LDPE chains are connected
to PS chains indicating that it is likely that not only anchoring
but also crosslinking of the soft particles should help to improve
the impact resistance. A similar investigation was also carried out
in the case of PE/PS blends in the presence of the previously described block and graft copolymers of well defined structure (see

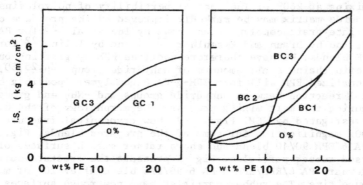

Figure 3.   Effect of 5% by weight added PS-PE copolymers on the
            impact strength of PS/PE blends at different compositions.
            PE content 0 – 25%. The structures of BC-1, BC-2 and BC-3
            block copolymers and those of GC-1, GC-3 and GC-4 graft
            copolymers are reported in Scheme 1.

Scheme 1) [13,15]. In Fig. 3 the effect of 5% added copolymer on the
impact strength is compared for different copolymers in the range
of 0 - 25% total LDPE. Anchoring appears to be adequate and, for PE
contents higher than 10%, reasonable values of the impact strength
resulted. In all cases electron microscopy revealed crazes formation
on the fracture surfaces. A different picture was found when impact
strength composition relationships were investigated in the region
of high contents of copolymer, up to 100% of the soft component.
The GC-1 and BC-1 containing blends have very small soft microphases
which are unable to stop catastrophic crazes and, consequently, show
a low impact strength. The blends containing partly or completely
microgelled GC-3 and GC-4 have particles of 1-10 μm and, accordingly,
better values of impact strength are found. The BC-2 and BC-3 copolymers form their own continuous phases in the blend and the fracture surfaces clearly show shear bands together with crazes. The
different effects on the impact strength caused by copolymers of
different structure are thus explained on the basis of different
deformation mechanisms.

Let us now consider Polyamide/Polyolefin blends. Aliphatic
polyamides have low impact resistance especially when dry and,
consequently, many attempts have been made to improve their impact
strength by proper blending with different polyolefins. However, most
of the results are reported in patent literature. In many of the
investigated PA/Polyolefin blends, graft copolymers are formed by
mechanochemical reactions promoted by shear forces [43-46]. However,
the use of polyolefins functionalized by grafting unsaturated anhy-
drides, dicarboxylic acids, or diesters onto the chain has also
been reported [47-49]. We have investigated blends of Polyamide 6
and of an amorphous ethylene/propylene copolymer (EPM) prepared by
melt mixing at 260° C. The poor dispersibility of polyolefins in
a polyamide matrix may be markedly improved by the presence of
appropriate graft copolymers as shown by Ide et al. [23] for PA 6/PP
blends, and by Braun and Eisenlhor [20,21] and by Illing [22] for
PA 6/PE blends. We have therefore modified EPM by grafting onto the
polyolefin chains small amounts of anhydride groups [24,25,50]. The
functionalized EPM will form EPM-g-PA 6 copolymers possibly by
chemical reaction between anhydride groups and $-NH_2$ end groups of
PA 6 during the melt mixing process. The morphology of the blends
was investigated by SEM. Figs. 4-7 show cryogenical fracture surfaces
at 5000x magnification for some of the prepared blends. Fig. 4 refers
to a PA 6/EPM 90/10 blend and shows rather smooth surfaces of the
dispersed rubber particles. Fig. 5 presents the fracture surface
of a ternary PA 6/EPM/EPM-g-PA 6 90/5/5 blend after 20' of mixing
residence time. The rubber particles have now rough surfaces and
matrix-dispersed phase connections are evident. These effects are
enhanced by increasing the mixing residence time which reasonably
favours the location of the copolymer at the rubber-polyamide interface, as illustrated in Fig. 6. Better adhesion is therefore the
result of the addition of modified EPM. Fig. 7 shows, finally at

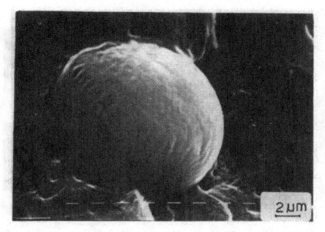

Figure 4.   Scanning electron micrograph of a cryogenic fracture
            surface of a PA 6/EPM 90/10 binary blend.

the same magnification as above, the cryogenical fracture surface
of a binary PA 6/EPM-g-PA 6 90/10 blend. No evidences of dispersed
rubber domains can be found, indicating a very high degree of homogenization in the absence of pure EPM. In order to investigate the
impact behaviour of the above reported blends, Izod impact tests
were performed at room temperature on compression molded specimens

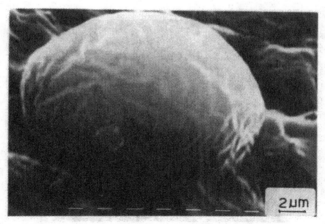

Figure 5.   Scanning electron micrograph of a cryogenic fracture
            surface of a PA 6/EPM/EPM-g-PA 6 90/5/5 ternary blend.
            Mixing residence time 20 min.

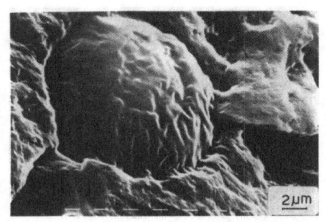

Figure 6.   Scanning electron micrograph of a cryogenic fracture
             of a PA 6/EPM/EPM-g-PA 6 90/5/5 ternary blends.
             Mixing residence time 60 min.

and the fracture surfaces were observed by SEM. In Figs. 8-10 are
shown the fracture surfaces in the induction zone at 320x magnification for some of the investigated blends. The binary PA 6/EPM blend
(Fig. 8) presents a morphology of the "cheesy" type with cavities
of about 20 μm formed by rubber particles pulled out during the
impact and characterized by clean and smooth surfaces. Replacement
of half of the EPM with the modified rubber (Fig. 9) leads to better

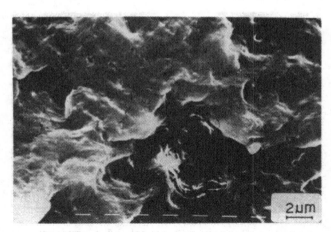

Figure 7.   Scanning electron micrograph of a cryogenic fracture
             surface of a PA 6/EPM-g-PA 6 90/10 binary blend.

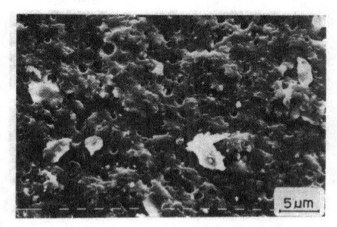

Figure 8. Scanning electron micrograph of an Izod impact test
fracture surface of a PA 6/EPM 90/10 binary blend,
Izod impact test performed at 20° C.

adhesion, as indicated by the presence of several rubber particles
anchored to the polyamide matrix. By increasing the mixing residence
time for this ternary blend, cavities of smaller and more uniform
size can be obtained (about 10 μm). Finally, when the soft component
is present exclusively as graft copolymer, no separate rubber domains

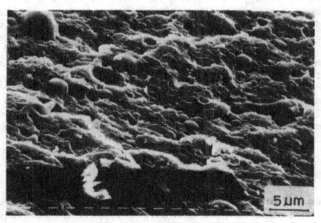

Figure 9. Scanning electron micrograph of an Izod impact test
fracture surface of a PA 6/EPM/EPM-g-PA 6 90/5/5 binary
blend. Izod test performed at 20° C.

Figure 10. Scanning electron micrograph of an Izod impact test
            fracture surface of a PA 6/EPM-g-PA 6 90/10 binary
            blend. Izod impact test performed at 20° C.

can be observed (Fig. 10). Moreover, the formation of tears and
fibrils on the fracture surface indicates that shear yielding contributes to the deformation mechanism under impact. In keeping with
this finding, ternary blends show an enhancement of the impact resistance with respect to EPM/PA 6 blends. The best results were found
in the case of binary PA 6/EPM-g-PA 6 blends where a more ductile
fracture mode seems to be operative. Although these results may be
considered preliminary, we can attempt to draw some conclusions on
the role of the graft copolymer in PA 6/EPM blends. In the ternary
blends it presumably acts as interfacial agent and impact property
improvements are achieved mostly by improved adhesion of rubber
particles; blends containing 100% graft copolymer as the soft component exhibit very fine dispersion and the marked impact resistance
improvement should be ascribed to modification of the fracture mode.
The overall picture of the investigated blends appears very complex
and further investigations seem necessary particularly to elucidate
the role of the graft copolymer.

CONCLUSIONS

    The use of interfacial agents in immiscible polymer blends is
of great importance if good mechanical properties are required. They
appear to be necessary especially when soft polymers are mixed with
rigid plastics in order to improve the impact resistance. At the
present time, however, general correlations between blend composition and molecular characteristics of the interfacial agent on the

one hand, and morphology and properties on the other hand have been
established only in a few cases.

ACKNOWLEDGMENTS

Many data reported in this paper have been obtained by our
colleagues of I.T.P.R. We gratefully acknowledge their kind cooperation. The authors also wish to acknowledge the financial support
of "Progetto Finalizzato Chimica Fine e secondaria del C.N.R.".

REFERENCES

1. N. A.J. Platzer ed., "Multicomponent Polymer Systems",
    Adv.Chem.Series, 99: Am.Chem.Soc., Washington (1971).
2. N. A.J. Platzer ed., "Copolymers, Polyblends and Composites"
    Adv.Chem.Series, 142: Am.Chem.Soc., Washington (1975).
3. L. H. Sperling ed., "Recent Advances in Polymer Blends, Grafts
    and Blocks", Plenum Press, New York (1974).
4. J. A. Manson and L.H. Sperling eds., "Polymer Blends and
    Composites", Plenum Press, New York (1976).
5. D. Klempner, K.C. Frish eds., "Polymer Alloys: Blends, Blocks,
    Grafts and Interpenetrating Polymer Networks", Plenum Press,
    New York, London (1977).
6. D. R. Paul and S. Newman eds., "Polymer Blends" vol. I, II,
    Academic Press, New York (1978).
7. E. Martuscelli, R. Palumbo and M. Kryszewski eds., "Polymer
    Blends; Processing, Morphology and Properties", Plenum
    Press, New York, London (1980).
8. D. R. Paul and J.W. Barlow, J.Macromol.Sci., Rev.Macromol.Chem.,
    C-18:109 (1980).
9. O. Olabisi, L.M. Robeson, M.T. Shaw, "Polymer-Polymer Miscibi-
    lity", Academic Press (1979).
10. N. G. Gaylord, in ref.[2], p. 76.
11. D. R. Paul, in ref.[6], p. 35, V. II.
12. A. Rudin, J.Macromol.Sci., Rev.Macromol.Chem., C-19:267 (1980).
13. D. Heikens, N. Hoen, W.M. Barentsen, P. Piet and H. Landan,
    J.Polym.Sci., Polymer Symposium, 62:309 (1978).
14. W. M. Barentsen, D. Heikens and P. Piet, Polymer, 15:122 (1974).
15. S. D. Sjoerdsma, A.C.A.M. Bleijenberg and D. Heikens, in ref.[7],
    p. 201
16. C. E. Locke and D.R. Paul, J.Appl.Polym.Sci., 17:2791 (1973).
17. G. Riess, in ref.[7], p. 123.
18. C. E. Locke and D.R. Paul, Polym.Eng.Sci., 13:308 (1973).
19. D. R. Paul, C.E. Locke and C.E. Vinson, Polym.Eng.Sci.,
    13:202 (1973).
20. D. Braun and V. Eisenlhor, Angew.Makromol.Chem., 58/59:22
    (1977).
21. D. Braun and U. Eisenlohr, Kunststoffe, 65:139 (1975).

22. G. Illing, Kunststoffe, 7:275 (1968).
23. F. Ide and A. Hasegawa, J.Appl.Polym.Sci., 18:963 (1974).
24. M. Avella, R. Greco, N. Lanzetta, G. Maglio, M. Malinconico,
    E. Martuscelli, R. Palumbo and G. Ragosta, in ref.[7], p. 191.
25. L. Amelino, S. Cimmino, R. Greco, N. Lanzetta, G. Maglio,
    M. Malinconico, E. Martuscelli, R. Palumbo and C. Silvestre,
    "Plasticon 81", Symposium on Polymer Blends, Warwich (1981).
26. G. E. Molau, J.Polym.Sci., Part A3:1267 (1965); Part A3:4235
    (1965).
27. G. Molau and W.M. Wittbrodt, Macromolecules, 1:260 (1968).
28. G. E. Molau, Kolloid Z. u. Z. Polymere, 238:439 (1970).
29. G. Riess, J. Periard and A. Bonderet, in "Colloidal and
    Morphological Behaviour of Block and Graft Copolymers",
    G.E. Molau ed., Plenum Press (1971).
30. T. Inoue, T. Soen, T. Hashimoto and H. Kawai, Macromolecules,
    3:87 (1970).
31. R. Fayt, R. Jerome, Ph. Teyssie, J.Polym.Sci., Polym.Lett.Ed.,
    19:79 (1981).
32. E. Nolley, J.W. Barlow and D.R. Paul, Polym.Eng.Sci., 20:364
    (1980).
33. L. D'Orazio, R. Greco, C. Mancarella, E. Martuscelli,
    G. Ragosta and C. Silvestre, (in press).
34. D. J. Meier, J.Polym.Sci. Part C, 26:81 (1969).
35. D. J. Meier, Am.Chem.Soc., Div.Polym.Chem.Prepr., 15:171 (1974).
36. L. Toy, M. Niinomi and M. Sghen, J.Macromol.Sci., Phys., B11:281
    (1975).
37. A. R. Ramos and R.E. Cohen, Polym.Eng.Sci., 17:639 (1977).
38. G. Riess, M. Schlienger and S. Marti, J.MacromolSci., Phys.,
    B-17:355 (1980).
39. G. C. Eastmond, M. Jiang and M. Malinconico, (in press).
40. G. Riess and Y. Jolivet, in ref.[2], p. 243.
41. G. C. Eastmond and D.G. Phillips, Polymer, 20:1501 (1979).
42. C. B. Bucknall, "Toughned Plastics", Applied Science Publ.LTD,
    London (1977).
43. Encyclopedia of Polymer Science and Technology, 2:503; 4:676,
    Interscience Publishers, New York (1965).
44. G. Illing, D. Zettler, Belg. Patent, 633,017 (1963).
45. H. Craubner, G. Illing, DBP, 1,131,883 (1962).
46. A. J. Chompff, U.S. Patent 3,880,948, (April, 29, 1975).
47. R. J. Kray and R.J. Bellet, French Patent, 1,470,255,
    (Febr. 17, 1967).
48. J. H. Davies, U.K. Patent 1,403,797 (1975).
49. E. I. Du Pont de Nemours, U.K. Patent 1,552,352, (1979).
50. M. Avella, N. Lanzetta, G. Maglio, M. Malinconico, P. Musto,
    R. Palumbo and M.G. Volpe, this book.

MORPHOLOGY, CRYSTALLIZATION AND MELTING BEHAVIOUR OF THIN FILMS

OF ISOTACTIC POLYPROPYLENE BLENDED WITH SEVERAL RUBBERS

E. Martuscelli, C. Silvestre and L. Bianchi

Istituto di Ricerca su Tecnologia
dei Polimeri e Reologia del CNR
Arco Felice (Napoli), Italy

INTRODUCTION

During the last several years, much effort has been spent on developing new materials, based on iPP/elastomers blends[1-4]. This interest is related to the fact that addition of the rubber phase improves the impact strength of the iPP. The present paper reports on a study of the isothermal crystallization and melting behaviour of thin films of isotactic polypropylene blended with an ethylene--propylene diene terpolymer and three samples of polyisobutylene with different molecular mass.

The main goal of the study was to enhance our understanding of how the blend composition, the chemical structure, and the molecular mass of the elastomer may influence the crystallization process of the iPP from the melt, the overall morphology of the blends, and the thermal behaviour of the semicrystalline matrix.

EXPERIMENTAL

The molecular characteristics, the source and the code of the polymer used in this study are listed in Table 1. Following purification of the starting polymers, binary blends of iPP/elastomers were prepared by evaporating the two components from xylene. Thin films were obtained by compression molding of the blend powder at about 200° C. The radial growth rate (G = dr/dt, r = radius of the spherulite, t = time) were calculated by measuring the size of iPP spherulites as a function of the time, using an optical polarizing microscope equipped with an automatized hot stage. The observed optical melting temperature, T'_m, of pure iPP and of iPP crystallized

Table 1. Molecular characteristics, source and code of the polymers
          used in the present investigation

Polymer	Source and trade name	Molecular mass	Viscosity money ML (1+4) 100° C
Isotactic Poly-propylene (iPP)	RAPRA	$\overline{M}_w = 3.07 \times 10^5$ $\overline{M}_n = 1.56 \times 10^4$	
Polyisobutylene (PiB)	Vistanex LM·MH (Esso) (PiB_{LM})	$\overline{M}_v = 6.6 \times 10^4$	
	Vistanex L120 (PiB_{MM})	$\overline{M}_v = 1.6 \times 10^6$	
	EGA Chemie (PiB_{HM})	$\overline{M}_v = 3.5 \times 10^6$	
Ethylene/Propy-lene/diene ter-polymer (EPDM)	Dutral TER 054-E Montedison		4

from the blend, was taken as the temperature at which the birefringence of the sample disappears. More details of the experimental
procedure used in this investigation can be found in a previous
paper[1].

RESULTS AND DISCUSSION

Overall Morphology

     An analysis of the optical micrographs of thin films of blends
shows that the overall morphology depends on the composition and on
the molecular mass of the elastomer.

     1) iPP/PiB$_{LM}$ blends: micrographs of films of such blends are
shown in Fig. 1. In the case of 90/10 blends, the elastomer is
ejected mainly at the growing spherulite boundaries. For the 80/20
blend the iPP spherulites are more open and coarse, suggesting that
part of the elastomer is probably incorporated in interlamellar
regions.

     2) iPP/PiB$_{MM}$ blends: in such blends (see Fig. 2) the elastomer

Fig. 1. Optical micrographs of melt crystallized films of
iPP/PiB$_{LM}$ blends, T$_C$ = 131° C, crossed polars;
a) 90/10, b) 80/20.

Fig. 2. Optical micrographs of melt crystallized films of iPP/PiB$_{MM}$ blends, T_c = 131° C, crossed polars; a) 90/10, b) 80/20, c) 60/40.

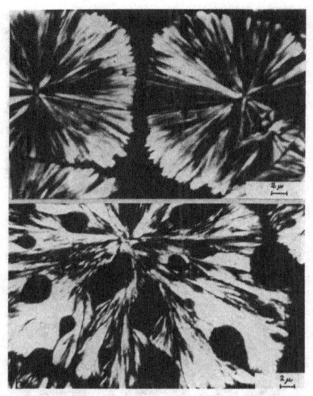

Fig. 3.   Optical micrographs of melt crystallized films of iPP/PiB$_{HM}$
           blends, T_C = 135° C, crossed polars;   a) 90/10,   b) 80/20.

separates from the iPP phase, forming spherical domains incorporated
in intraspherulitic regions.

      3)   iPP/PiB$_{HM}$ blends: in films of iPP/PiB$_{HM}$ 90/10 blend (see
Fig. 3) the elastomer is mainly ejected at the spherulites boundaries. In the case of 80/20 blend, spherical elastomeric domains
in both intra- and interspherulitic regions are observed. Inter-
conneted morphology is found in blends with higher PiB$_{HM}$ content.

      4)   iPP/EPDM blends: optical micrographs, presented in Fig. 4,
show that the elastomer separates in droplet-like domains mainly
dispersed in intraspherulitic regions. These rubbery domains seem
to be aligned along the radial direction.

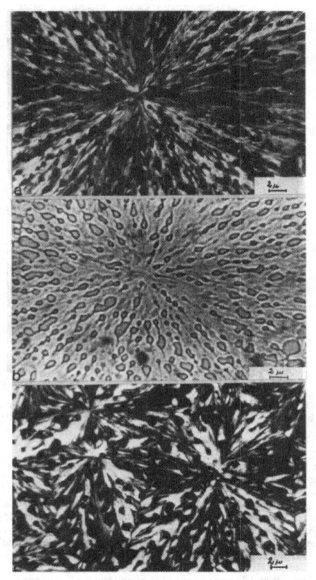

Fig. 4.   Optical micrographs of melt crystallized films of iPP/EPDM
           blends, T_c = 135° C;   a) 90/10, crossed polars,   b) 90/10,
           parallel polars,   c) 60/40, crossed polars.

Nucleation density

The addition of rubber to the iPP gives rise to differences in the nucleation density of the spherulites. The number of spherulites per unit area (N/S) depends on the chemical nature and molecular mass of the elastomer, on T_c and composition. For iPP/EPDM blends, N/S

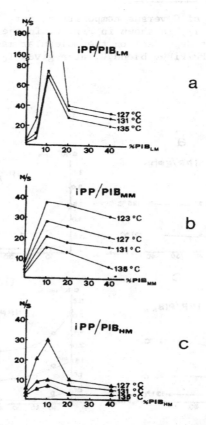

Fig. 5. Number of spherulites per unit area, N/S, vs. elastomer content, at different T_c: a) iPP/PiB$_{LM}$ blends; b) iPP/PiB$_{MM}$ blends; c) iPP/PiB$_{HM}$ blends.

increases with increasing rubber content, whereas for the iPP/PiB
blends, plots of N/S against the elastomer content exhibit a maximum
for a composition of about 10% (see Fig. 5).

From the results reported above it may be concluded that the
addition of rubber to iPP produces drastic changes on the shape,
dimension, and structure of the spherulite, and of the interspherulitic structure. It is well known that these factors have a great
influence on the crack propagation and should be taken into account
to explain the fracture mechanism of the iPP/rubber blends[5].

Spherulite radial growth rate

The variation of G versus composition, for iPP/EPDM and iPP/PiB
blends, at a given T_c, is shown in Fig. 6. In the case of iPP/PiB
blends the trend is dependent on the molecular mass of PiB. It can
be seen that for iPP/PiB$_{MM}$ blend G, at a given T_c, decreases

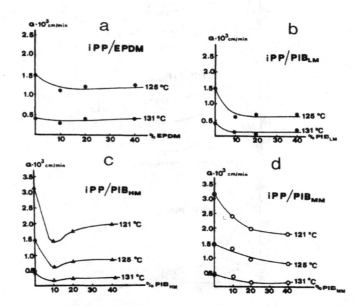

Fig. 6.   Radial growth rate, $G \times 10^3$ (cm/min), vs. elastomer content
         at different T_c: a) iPP/EPDM blends, b) iPP/PiB$_{LM}$ blends,
         c) iPP/PiB$_{MM}$ blends, d) iPP/PiB$_{HM}$ blends.

monotonically with the elastomer content, whereas plots of G versus composition for iPP/PiB$_{HM}$ show a minimum at 90/10 blend composition. In the case of iPP/PiB$_{LM}$ after an initial drop, G remains almost constant as the rubber content increases. Plots of G versus % of elastomer in the case of iPP/PiB$_{MM}$, iPP/PiB$_{HM}$ at ΔT = constant, show maxima (see Fig. 7) more pronounced at higher ΔT, for an elastomer content of about 20%. By contrast, a minimum for 10% of rubber is observed in G $\rightarrow$ % of the elastomer plots of iPP/PiB$_{LM}$ blends.

For a polymer/diluent system the temperature dependence of G may be accounted for in terms of the following equation[6],[7]

$$\text{LogG} - \text{Log}\nu_2 + \frac{\Delta F^*}{2.3RT_c} - \frac{2 \sigma T_m \text{Log}\nu_2}{b_o \Delta H_v \ \Delta T} =$$

$$= \text{LogG}_o - \frac{4b_o \sigma \sigma_e}{2.3K \Delta H \Delta T} \cdot \frac{T_m}{T_c} \tag{1}$$

Eq. 1 fits quite well with the experimental data for the iPP/PiB and iPP/EPDM blends. From the intercept and the slope it was possible to calculate, for every composition, the values of the folding surface free energy σ_e and of LogG$_o$. Note that in the calculation of ΔF^*, the T_g of the blend was assumed to be equal to that of pure iPP (T_g = -18° C), and the volume fractions were calculated by using density values measured at 180° C (ℓ_{iPP} = 0.766 g/cm^3, ref.[8], ℓ_{PiB} = = 0.839 g/cm^3, ref.[9]).

Melting behaviour

The T'$_m$ of iPP crystallized from iPP/elastomers mixtures increases linearly with T_c, according to the relation of Hoffman[10]. At a given T_c the observed melting temperatures of the blends are always lower than that of the pure iPP. In particular, for iPP/PiB$_{LM}$ and iPP/PiB$_{MM}$ blends, T'$_m$ decreases monotonically with the elastomer content, whereas a minimum is observed in the iPP/PiB$_{HM}$ blends at the 90/10 composition (Fig. 8). The variation of the equilibrium melting temperature, T_m, obtained from the Hoffman equation, with the elastomer content is shown in Fig. 9.

For iPP/PiB$_{MM}$ and iPP/PiB$_{HM}$ the T_m $\rightarrow$ % of elastomer curves present a minimum for the 80/20 composition (see Fig. 10). In the case of iPP/PiB$_{LM}$ blends the T_m $\rightarrow$ % PiB curve presents a maximum at a 90/10 composition. Blends with higher PiB$_{LM}$ content (20%) have positive values of the equilibrium melting point depression ΔT^o_m, (T^o_m (iPP) $- T_m$ blend)). In the case of iPP/EPDM blends we observed a minimum in the T_m $\rightarrow$ % EPDM plot for a 80/20 composition (see Fig. 9). Thus iPP/EPDM blends behave qualitatively as iPP/PiB$_{HM}$ blends.

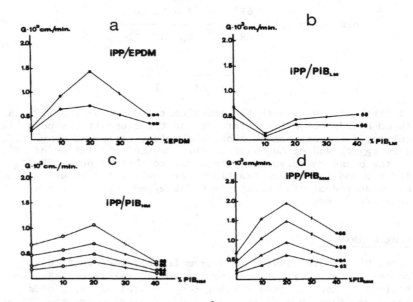

Fig. 7.   Radial growth rate, $G \times 10^3$ (cm/min), vs. elastomer content at different undercooling $\Delta T(T_m - T_c)$:   a) iPP/EPDM blends;   b) iPP/PiB$_{LM}$ blends;   c) iPP/PiB$_{MM}$ blends,   d) iPP/PiB$_{HM}$ blends.

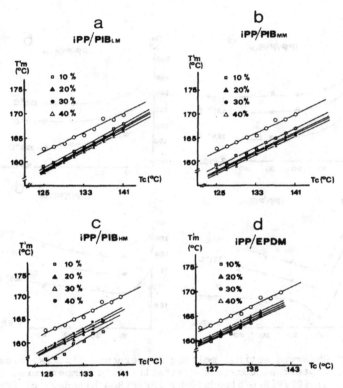

Fig. 8. Variation of the observed optical melting temperature T'_m, with crystallization temperature, T_c, at constant composition: a) iPP and iPP/PiB$_{LM}$ blends; b) iPP and iPP/PiB$_{MM}$ blends; c) iPP and iPP/PiB$_{HM}$ blends; d) iPP and iPP/EPDM blends.

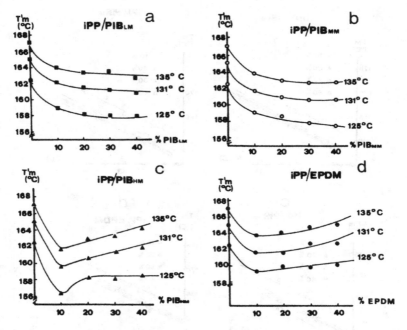

Fig. 9. Observed optical melting temperature T'_m, vs. elastomer
content at constant crystallization temperature, T_c:
a) iPP/PiB$_{LM}$ blends, b) iPP/PiB$_{MM}$ blends, c) iPP/PiB$_{HM}$
blends and   d) iPP/EPDM blends.

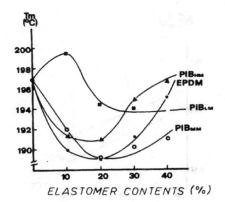

Fig. 10. Variation of equilibrium melting temperature, T_m with
the elastomer percentage.

Such behaviours may be probably accounted for if:
1) PiB_{MM}, PiB_{HM} and EPDM are able to act as diluents for iPP at
lower concentration. In blends with elastomer content larger
than 20%, the mutual solubility of the two components in the
melt, decreases with the increase of the elastomer content;
2) in the case of iPP/PIB_{LM} blends the maximum in T_m is probably
related to the fact that the PiB_{LM} is able to selectively
dissolve a certain amount of the more defective iPP molecules
by analogy with what we found for iPP/EPM copolymers blends[1].
At a higher PiB_{LM} content, T^o_m becomes positive presumably
because the diluent effects prevail. Kinetic effects,
moreover, could also be present.

CONCLUSIONS

The type of dependences on the composition, on the crystallization temperature, and on the chemical nature and molecular mass of
the components observed in kinetic and thermodynamic properties,
relative to the isothermal crystallization process, the final overall
morphology, and the thermal behaviour, are all to be related to the
physical state of the melt, which at T_c is in equilibrium with the
developing solid phase.

The trend of T_m and G versus composition for the iPP/PiB and
iPP/EPDM blends suggests that the polymers are semicompatible at
the investigated T_c. This hypothesis is in agreement with the
finding that the solubility parameters δ, calculated[11] using Hoy's
tables[12] for the molar atraction constans and literatura data[8,9]

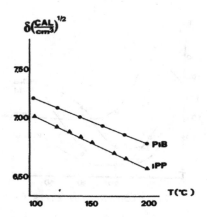

Fig. 11.   Variation of solubility parameter, δ $(cal/cm^3)^{1/2}$
with the temperature for iPP and PiB.

for the density values of the iPP and PiB, are very close (see
Fig. 11).

The minima observed in plots of T'_m and T_m against the content
of the uncrystallizable component in iPP/PiB$_{HM}$ and iPP/EPDM systems
and in plots of G against composition at constant T_C are probably
related to processes of phase separation, followed by molecular
fractionation and preferential dissolution of smaller and/or more
defective molecules of the crystallizable component in the domains
of the uncrystallized polymer.

REFERENCES

1. E. Martuscelli, C. Silvestre and G.C. Abate, Polymer, 23:229
    (1982).
2. S. Danesi and R.S. Porter, Polymer, 19:448 (1978).
3. G. A. Ermidowa, I.A. Ragozina and N.M. Leont'eva, Plast.Massy,
    5:52 (1969).
4. K. J. Kumbhoni, Polymer Progress Plast.Ed. , 3 (1974).
5. K. Friedrich, Progress in Colloid and Polymer Science, 64:103
    (1978).
6. D. Turnbull and J.C. Fisher, J.Chem.Phys., 17:71 (1949).
7. J. Boon and J.M. Azcue, J.Polym.Sci., A-2, 6:885 (1968).
8. F. Danusso, G. Moraglio, W. Ghiglia, L. Motta and G. Todomini,
    Chim.Ind., 41:749 (1959).
9. B. E. Eichonger and P.J. Flory, Macromolecules, 1:285 (1968).
10. J. D. Hoffmann, SPE Trans, 4:315 (1964).

11. S. Krause in "Polymer Blends" D.R. Paul and S. Newman eds.,
      Chapt. 2, Academic Press, New York (1978).
12. K. L. J. Hoy, Paint Technol, 42:76 (1970).

This work was partialy supported by "Progetto Finalizatoo
Chimica Fine e secondaria" del CNR.

PROPERTIES OF POLY(ETHYLENE OXIDE) / POLY(METHYLMETHACRYLATE)

BLENDS: MORPHOLOGY, CRYSTALLIZATION AND MELTING BEHAVIOUR

E. Martuscelli[a], M. Pracella[b], and Wang Ping Yue[c]

[a] Istituto di Ricerche su Tecnologia dei Polimeri
e Reologia del C.N.R., Arco Felice NA , Italy

[b] Centro Studi Processi di Polimerizzazione e Pro-
prieta Fisiche e Tecnologiche dei Sistemi Macromo-
lecolari del C.N.R., Dip.Ing.Chim., Pisa, Italy

[c] North Western Chemical Power Corp., Xian, China

INTRODUCTION

Blends of low molecular weight poly(ethylene oxide) (PEO) with poly(methylmethacrylate) (PMMA) show unique morphology, depression of spherulite growth rate, and melting temperature of PEO with increasing PMMA concentration, a glass transition temperature inter-mediate between those of pure components[1]. This and other evidence, based on the dynamical-mechanical properties, light scattering and electron microscopy investigation[2], indicate that PEO/PMMA blends are compatible in the molten state.

In the present paper the properties of blends of high molecular weight PEO with high molecular weight PMMA are reported. The main aim of this research is to analyze the influence of composition and mole-cular weight of the components in PEO/PMMA blends on the morphology, on the melting process and on the kinetic and thermodynamic parame-ters controlling the spherulite growth rate and the overall rate of crystallization from the melt.

EXPERIMENTAL

Blends of PEO with $M_w = 1 \times 10^5$ (Fluka A.G.) and PMMA with $M_w = 1.1 \times 10^5$ (BDH) in a composition range from 100% to 60% of PEO were prepared by solution casting from chloroform and then

drying under vacuum at 80° C for 24 hours to remove the residual
solvent.

The morphology and the isothermal radial growth rate of PEO
spherulites in the blends were studied on thin films of these samples
using a Reichert polarizing microscope equipped with a Mettler hot
stage. The films were first melted at 85° C for 5 minutes, following
which they were rapidly cooled to a fixed crystallization temperature T_c and the radius of the growing spherulites was measured as
a function of time.

The overall crystallization kinetics of molten blends were
analyzed by differential scanning calorimetry with a Perkin-Elmer
DSC 2 apparatus. Following melting, the samples were heated at 85° C
for 5 min. and isothermally crystallized at various T_c recording
the heat of crystallization as a function of permanence time. The
fraction X_t of the material crystallized after time t was determined
by means of the relation $X_t = Q_t/Q_\infty$, where Q_t is the heat generated
at time t and Q_∞ is the total heat of crystallization for t = ∞.

The melting temperatures T_m' of the isothermally crystallized
blends were measured both by DSC and optical microscopy by heating
the samples directly from T_c to T_m with heating rates of 20° and
1°/min. respectively. The mass crystallinity X_c of the blends was
calculated for various T_c by the ratio between the apparent enthalpy
of fusion of DSC endotherms and the enthalpy of fusion of 100%
crystalline PEO.

RESULTS

Morphology and spherulitic growth rate

Films of PEO/PMMA blends isothermally crystallized in the temperature interval 39°-56° C show in the composition range examined
a spherulitic morphology with the characteristic birefringent pattern
in the form of Maltese cross when observed between crossed polaroids.
No segregation phenomena of the noncrystallizable component in the
intraspherulitic regions have been observed. With decreasing PEO
content in the blends the texture of the spherulites becomes pro-
gressively more open and coarse (Fig. 1). Moreover, after crystallization the films appear completely filled with impinged spherulites.
For blends with a PEO content of less than 60%, non-impinging spherulites and rod-like crystals have been observed[2]. The average dimensions of the spherulites in the blends change to some extent with
PMMA concentration.

For each T_c the spherulite radius R increases linearly with time
t and no decrease of the growth rate G = dR/dt is observed over long
time indicating that during the growth the concentration   of

Fig. 1. Optical micrographs (crossed polaroids) of spherulites growing in thin films of PEO/PMMA blends of various compositions: a) PEO/PMMA 90/10, $T_c = 53°$ C; b) PEO/PMMA 80/20, $T_c = 48°$ C; c) PEO/PMMA 70/30, $T_c = 48°$ C; d) PEO/PMMA 60/40, $T_c = 46°$ C.

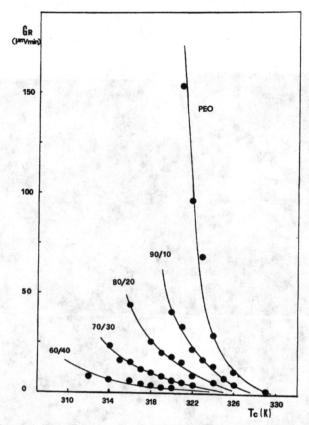

Fig. 2.   Radial growth rate G of spherulites in pure PEO and PEO/PMMA
          blends as a function of crystallization temperature.

non-crystallizing material at the tips of the radial lamellae is
constant. The temperature dependence of G with varying blend composition indicates that the presence of PMMA induces a depression of
the spherulite growth rate the more the higher is the concentration
of non-crystallizing polymer and the lower is the crystallization
temperature (Fig. 2).

      Similar behaviour has also been observed in the case of
poly(caprolactone)/poly(vinyl chloride) blends[3] and poly(vinyli-
dene fluoride)/poly(methylmethacrylate) blends[4]. These results
clearly suggest that during crystallization PMMA is incorporated
in the interlamellar regions of PEO spherulites.

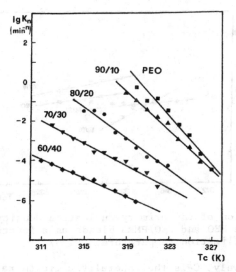

Fig. 3. Temperature dependence of the overall kinetic constant K_n (see Eq. (1)) for various PEO/PMMA blends isothermally crystallized from the melt.

Overall crystallization kinetics

The isothermal crystallization kinetics from the melt of all PEO/PMMA blends follow the Avrami equation until a high degree of conversion. The values of the overall kinetic constant K_n and of the Avrami index n have been calculated at each T_c in accordance with the expression:

$$X_t = 1 - \exp(-K_n t^n) \tag{1}$$

For the same T_c, K_n strongly decreases with increasing PMMA content while the average value of the index n varies from 2.5 to 3.1 similarly as in the case of crystallization process proceeding by heterogeneous nucleation and threedimensional growth of the crystals (Fig. 3).

The variation of the overall crystallization rate has thus been related to the nucleation density in the melt by using the relation[5]:

$$K_n = \frac{4\pi \rho_c}{3\rho_a (1-\lambda_\infty)} G^3 N \tag{2}$$

where ρ_c and ρ_a are the densities of the crystalline and amorphous

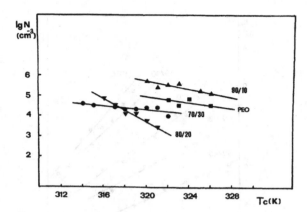

Fig. 4. Variation of the primary nucleation density N (see eq. (b))
         in pure PEO and PEO/PMMA blends as a function of the
         crystallization temperature.

phases respectively, G is the spherulitic growth rate, N is the
number of potential nuclei per unit volume, and $(1-\lambda_\infty)$ is the weight
fraction of polymer that is crystalline at $t = \infty$. For all the examined
blends the values of N decrease with increasing T_c and, for the same
T_c, depend on the blend composition (Fig. 4). Under the same undercooling, the value of N in the PEO/PMMA 90/10 blend is about ten
times larger than that for pure PEO while for the 70/30 and 80/20
blends it is slightly suppressed. In accordance with this trend,
spherulites from the 90/10 blend have smaller sizes than in the
other samples.

Melting behaviour

        For all the blends the PEO melting temperature T_m' linearly
increases with T_c following the equation[6]:

$$T_m' = (\gamma-1)\, T_m/\gamma + T_c/\gamma \,, \qquad\qquad (3)$$

where T_m is the equilibrium melting temperature and γ is a parameter
depending on the thickening phenomena of the lamellae at T_c. For the
same T_c the melting temperature T_m' of the blends decreases with
increasing PMMA content, particularly at high values of the undercoolings explored. At lower undercoolings, the melting points of each
blend undergo an abrupt increase and approach the values of T_m of
pure PEO.

        The slope of the linear trends is independent of blend composi-
tion suggesting that the depression of the melting points is to be

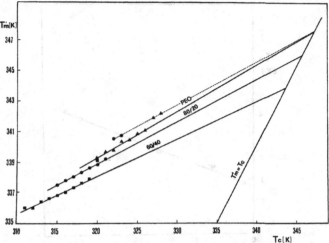

Fig. 5.   Observed melting temperatures T_m' (DSC measurements) of PEO
and its blends with PMMA after isothermal crystallization
at various T_c. The extrapolation of the linear plots of T_m'
to the line $T_m' = T_c$ gives the values of the equilibrium
melting temperature T_m.

ascribed mainly to the diluent effect of PMMA in the molten blends.
The nonlinear variation of T_m' at higher T_c could be accounted for by
a decrease of solubility of the two polymers in the melt as the temperature increases (Lower Critical Solution Temperature Behaviour)
(Fig. 5).

The values of the equilibrium melting temperature T_m, determined
from Eq. (3), decrease in a continuous way with increasing PMMA content (Fig. 6). The depression observed for the equilibrium melting
temperature of pure PEO, $\Delta T_m = T_m^o - T_m$ (where T_m^o refers to pure PEO)
has then been related to the volume fraction ν_1 of the non-crystallizing polymer according to the equation derived by Nishi and Wang[7]
for compatible binary blends:

$$\Delta T_m = -T_m^o \left(\frac{V_{2u}}{\Delta H_{2u}}\right) B\nu_1^2 \qquad (4)$$

where $\Delta H_{2u}/V_{2u}$ is the latent heat of fusion per unit volume, and
$B = RT\ X_{12}/V_{1u}$ is determined from the molar volume of the noncrystallizing component V_{1u} and from the Flory-Huggins interaction parameter
X_{12}. From the slope of the plot of ΔT_m versus ν_1^2, a value of -2.85
was calculated for B.

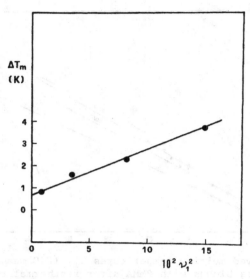

Fig. 6. Equilibrium melting point depression of PEO/PMMA blends
(see Eq. (4)) as a function of the volume fraction of PMMA.

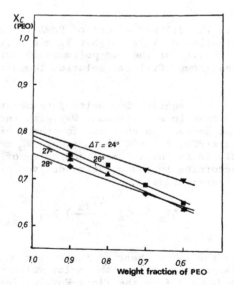

Fig. 7. Variation of the crystallinity fraction of PEO in the
blends for different values of the undercooling.

This value is close to that found for other compatible polymer blends[7-9].

The crystallinity fraction X_c(PEO) of the crystalline component in the blends is also influenced by the composition. At constant undercooling, X_c(PEO) becomes a decreasing function of the PMMA weight fraction (Fig. 7).

Temperature dependence of G and K_n

The influence of PMMA content on the kinetic and thermodynamic parameters controlling the isothermal spherulitic growth and the overall crystallization rate of PEO from the molten blends has been analyzed on the basis of the modified Turnbull-Fisher equation[10]:

$$\ln G = \ln(\nu_2 \, G_o) - \frac{\Delta F_{WLF}}{RT_c} - \frac{\Delta \Phi^*}{KT_c} \quad , \tag{5}$$

where ν_2 is the PEO volume fraction, ΔF_{WLF} - the activation energy for the transport of crystallizing units across the liquid-solid interface, $\Delta \Phi^*$ - the free energy required to form a nucleus of critical size, and K is the Boltzmann constant. For a polymer-diluent system, $\Delta \Phi^*$ is expressed as:

$$\Delta \Phi^* = \frac{4b_o \, \sigma \, \sigma_e \, T_m}{\Delta H_f(T_m - T_c)} - \frac{2 \, \sigma \, K \, T_c \, \ln \nu_2}{b_o \, \Delta H_f(T_m - T_c)} \quad , \tag{6}$$

where b_o is the distance between adjacent fold planes, σ and σ_e are the interfacial free energies per unit area parallel and perpendicular to molecular chain axis respectively, and ΔH_f is the fusion enthalpy of PEO per unit volume[1].

By plotting the quantity $f(G) = \ln G - \ln \nu_2 + (\Delta F_{WLF}/RT_c) - (0.2T_m \ln \nu_2)/(T_m - T_c)$ against $T_m/T_c \, (T_m - T_c)$ for the various blends, straight lines are obtained from whose slopes it is possible to calculate the values of the free energy of folding σ_e (Figs. 8-9). A notable decrease of σ_e has been found for increasing concentrations of PMMA in the blends indicating that probably during the crystallization the molecules of the noncrystallizing component can be trapped in the interlamellar regions of the PEO spherulites causing perturbations in the structure of the crystal surface.

Finally, the crystallization process of PEO/PMMA blends is influenced by the PEO molecular weight. Both the variation of the spherulitic growth rate, melting temperature and energy of formation of surface nuclei with the composition indicate that the molecular weight plays a decisive role mainly at high values of the

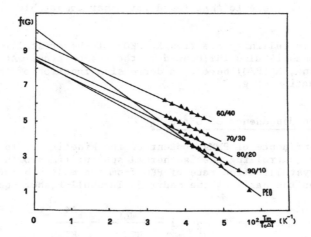

Fig. 8. Plots of the quantity f(G) against T_m/T_c ΔT according to Eqs. (5) and (6) for PEO and PMMA blends.

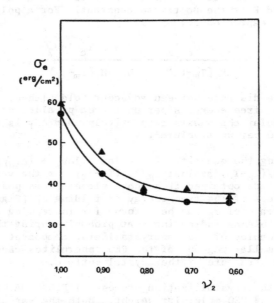

Fig. 9. Variation of the free energy of folding σ_e with the volume fraction of PEO (▲ : values calculated from K_n plots; ● : values calculated from G plots).

undercooling, while for lower values the crystallization process is controlled by the amount of PMMA. Differences in miscibility of the two components at low PMMA contents and variations of molecular transport in the melt could be accounted for by the different behaviour observed.

REFERENCES

1. E. Martuscelli and G.B. Demma, in "Polymer Blends: Processing, Morphology and Properties", E. Martuscelli, R. Palumbo and M. Kryszewski eds., Plenum Press, New York (1980).
2. D. F. Hoffmann, Ph.D. Thesis, University of Massachussetts (1979).
3. C. J. Ong and F.P. Price, J.Polym.Sci., Polym.Symp., 63:45, (1978).
4. T. Wang and T. Nishi, Macromolecules, 10:421 (1977).
5. L. Mandelkern, "Crystallization of Polymers", McGraw Hill, New York (1964).
6 J. D. Hoffmann, SPE Trans., 4:315 (1964).
7. T. Nishi and T. Wang, Macromolecules, 8:909 (1975).
8. R. L. Imken, D.R. Paul, J.W. Borlow, Polym.Eng.Sci., 16:593, (1976).
9. T. K. Kwei, G.D. Patterson, T. Wang, Macromolecules, 9:780, (1976).
10. J. Boon and J.M. Azcue, J.Polym.Sci. Part A2, 6:885 (1968).

SUSPENSION MODEL FOR ELASTIC MODULUS

OF TWO-COMPONENT POLYMER BLENDS

Z. Dobkowski

Institute of Industrial Chemistry
01-793 Warsaw, Poland

Reinforcement factor $e=(E/E_2)-1$, where E and E_2 are the moduli of blend and polymer matrix, respectively, as a function of volume fraction of the reinforcing material (v) is proposed for treatment of experimental data, as well as for comparison of different theoretical models for the elastic modulus of polymer blends.

The suspension model of Einstein and Guth is extended into 4-th order function of v, and the parallel voids model is extended into the 3-rd order function of v. The fraction of reinforcing material in the filler space ξ_R is considered as a measure of the efficiency of the reinforcing material for a given two-component polymer system, indicating also the state of adhesion at the phase boundary. The efficiency of a given blending method can this be estimated and compared with other methods. Moreover, the average interphase thickness can be calculated from the ξ_R value.

Examples based on published results are given for polypropylene-polycarbonate (PP-PC) and polycarbonate -polymethylmethacrylate (PC-PMMA) blends. For PP-PC non -compatible blends the average interphase thickness of about 0.05-0.30 μm is found.

INTRODUCTION

Improvement of the mechanical properties is of major importance for polymer blending. It is often useful to represent this improvement by the increase of elastic modulus, e.g. of Young's (tensile) modulus E. The elastic modulus E of two-component blend depends, above all, on the moduli of the components, i.e. of reinforcing material E_1 and of polymer matrix E_2, and on the volume fraction of the reinforcing material v. Other factors, however, such as the degree of compatibility (e.g. the interaction between the components), morphology (e.g. the size and shape of domains), Poisson's ratio, etc., play an important role. Many equations describing theoretical models for the calculation of Young's modulus of polymer blends have been reported and compared with experimental data [1,2]. Here a novel approach to this problem is described and discussed, based essentially on the known suspension model. Some examples are also given.

REINFORCEMENT FUNCTION

It has recently been found [3] that the theoretical models can conveniently be analysed using a function, called here the reinforcement function,

$$e = f \ (v, \ r_E, \ ...), \tag{1}$$

where $e=E/E_2 -1$ is the reinforcement factor, $r_E=E_1/E_2$ is the ratio of moduli of pure components, and the boundary conditions are $e=0$ or $E=E_2$ for $v=0$, and $e=r_E-1$ or $E=E_1$ for $v=1$.

Several models have been compared in this way (Table 1). For example, simple theoretical models can be expressed in terms of the following equations:
the parallel (additive) model, for equal strains,

$$e = (r_E - 1) \ v, \tag{2}$$

and the series model, for equal stresses,

$$e = (r_E - 1) \ v \ / \ [r_E-(r_E-1)v]. \tag{3}$$

Usually, the parallel and series models are considered to form the upper and lower bounds respectively, and the experimental data fall within these bounds [1]. We suggest that all models which express the dependence of E vs. v can be written as

$$e = (r_E - 1)v\phi. \tag{4}$$

The factor ϕ in eq. (4) depends, as it is shown in Table 1, on the volume fraction of the reinforcing material v, the ratio of moduli r_E, Poisson's ratio ν, the shape factor of the reinforcing material expressed by the length-to-diameter ratio l/d, the maximum possible packing fraction for a hard filler v_m, the voids fraction in the

Table 1. Reinforcement function $e = (r_E-1)v\phi$ for theoretical models of two-component systems

No.	Theoretical model			Factor ϕ this work, eq. (4)
	Model	Equation	Ref.	
1	Simple, parallel	$E=vE_1+(1-v)E_2$	1	1
2	Simple, series	$1/E=v/E_1+(1-v)/E_2$	1	$1/\left[r_E-(r_E-1)v\right]$
3	Kerner, 1956	$\dfrac{E}{E_2} = \dfrac{(\alpha+v)E_1 + (1-v)E_2}{\alpha(1-v)E_1+(1+\alpha v)E_2}$ $\alpha = (8-10\nu)/(7-5\nu),\ \nu$ is the Poisson's ratio of matrix	1,2,4	$\dfrac{1 + \alpha}{1+\alpha\left[r_E-(r_E-1)v\right]}$
4	Tsai-Halpin, 1968	$E/E_2=(1+ABv)/(1-Bv)$ $A = f(\nu,\ 1/d)$ $B = \left[(E_1/E_2)-1\right]/\left[(E_1/E_2)+A\right]$	1	$\dfrac{1 + A}{A+r_E-(r_E-1)v}$
5	Nielsen-Lewis, 1969	$E/E_2=(1+ABv)/(1-\psi Bv)$ $\psi = f(v,v_m),\ v_m$ is the maximum packing fraction $A = (7-5\nu)/(8-10\nu)$ $B = \left[(E_1/E_2)-1\right]/\left[(E_1/E_2)+A\right]$	1	$\dfrac{\psi + A}{A+r_E-(r_E-1)v}$
6	Halpin-Kardos, 1972	$E/E_2=(1+sBv)/(1-Bv)$ $s = 2(1/d)$ $B = \left[(E_1/E_2)-1\right]/\left[(E_1/E_2)+s\right]$	5	$\dfrac{1 + s}{s+r_E-(r_E-1)v}$
7	Parallel voids	$E = \xi_R vE_1+(1-v)E_2$ ξ_R is the fraction of reinforcing material in the filler space	3	$\dfrac{\xi_R r_E-1}{r_E-1}$

E, E_1 and E_2 are the moduli of blend, reinforcing material and polymer matrix, respectively; $r_E = E_1/E_2$; v is the volume fraction of reinforcing material.

filler space ξ_L, etc. However, it is often impossible to determine all the magnitudes influencing the factor, especially for polymeric systems. Moreover, experimental data for polymeric systems agree with theoretical models only in isolated cases.

SUSPENSION MODEL

It seems then that the most convenient model for polymeric systems is the suspension model of Einstein, cf. [1], extended by Guth [1,2,6] into the following form

$$E/E_2 = 1 + a_1 v + a_2 v^2, \tag{5}$$

which is often used for systems in the rubbery state. Eq. (5) can further be extended, and expressed in terms of the reinforcement factor e:

$$e = a_1 v + a_2 v^2 + a_3 v^3 + a_4 v^4 + \ldots \tag{6}$$

The coefficients of eq. (6) for a given two-component system can be calculated from experimental data. It has been found [3] that the first term of eq. (6) should have the form of a simple parallel (additive) model, and higher order terms express deviations from the rule of additivity. Thus, eq. (6) can be written as

$$e = (r_E - 1)v \left[k_1 + (1-k_1)v \, S_1/S_2 \right], \tag{7}$$

where k_1 expresses deviations from the simple parallel (additive) model,

$$S_1 = \sum_{i=o}^{n} p_i v^i \tag{7a}$$

$$S_2 = \sum_{i=o}^{n} p_i \tag{}$$

and p_i is the auxiliary parameter determined experimentally as the ratio of coefficients

$$p_i = a_{i+2}/a_2. \tag{7c}$$

It can also be written that

$$\frac{S_1}{S_2} = \frac{a_2 + a_3 v + a_4 v^2 + \ldots}{a_2 + a_3 + a_4 + \ldots}, \tag{7d}$$

if at least $a_2 \neq 0$. Eq. (7) is consistent with eq. (4) where

$$\phi = k_1 + (1-k_1)v\, S_1/S_2. \tag{7e}$$

The boundary conditions are also satisfied, i.e. e=0 or $E=E_2$ for v=0 and $e=r_E-1$, or $E=E_1$ for v=1. For $k_1=1$, eq. (7) becames identical with eq. (2), which expresses the rule of additivity.

Some equations for hypothetical polymeric systems with $r_E=2$ are compared in Fig. 1 as an example. For the suspension model, equations up to the 4-th order have been taken into account. It is evident that systems with modulus higher than expected from the parallel model can be described by the suspension model.

EXAMPLES OF APPLICATION

1. Polypropylene - polycarbonate (PP-PC) blends.

The tensile modulus E of PP-PC blends, prepared by mechanical

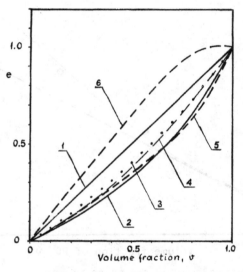

Figure 1. Comparison of different models for elastic modulus, $r_E=2$.
1. the parallel model; 2. the series model; 3. the Kerner and Nielsen-Lewis models for $v=0.35$ and the Halpin-Kardos model for s=1; 4. the suspension model $k_1=0.5$, $p_1=0$, $p_2=0$; 5. the suspension model $k_1=0.5$, $p_1=-2$, $p_2=1.5$; 6. the suspension model $k_1=1.5$, $p_1=-2$, $p_2=1.5$.

mixing of pellets in an extruder, was measured according to Polish
Standard PN 76/C-89051 using an Instron instrument [7,8]. The reinforcement factor e was then calculated and the results are shown in
Table 2 and Fig. 2. It is evident that the rule of additivity (or
the parallel model) is not satisfied, nor are equations of 2-nd and
3-rd order for the suspension model. It has been found that a 4-th
order equation fits the experimental data well and the following
regression equation has been obtained by the least-squares method:

$$e = 1.667v - 4.324v^2 + 10.245v^3 - 6.479v^4 \qquad (8)$$

with the coefficient of multiple correlation equal to 0.89. The plot
of eq. (8) is shown in Fig. 2. Coefficient k_1 can thus be calculated.
Comparing eqs. (6) and (7), we have

$$k_1 = \frac{a_1}{r_E - 1}, \qquad (9a)$$

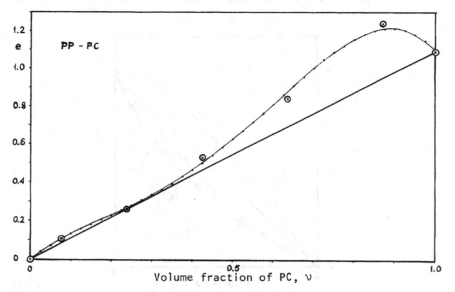

Figure 2. Plot of e vs. v for PP-PC blends.
⊙ experimental data [7,8], ··· the 4th order equation
for the suspension model;
— the parallel model.

Table 2.   Data for PP-PC blends

	Experimental data			Calculations	
Concentration of PC wt. %	Modulus $E \times 10^{-3}$ MPa	Density g/cm^3	Volume fraction of PC v	e	ξ_R
0	1.03	0.890	0	0	–
10	1.14	0.913	0.076	0.1068	1.1469
30	1.30	0.966	0.241	0.2621	0.9954
50	1.58	1.011	0.426	0.5340	1.0746
70	1.90	1.063	0.634	0.8447	1.1122
90	2.31	1.132	0.870	1.2427	1.1580
100	2.16	1.200	1	1.0971	1

or

$$k_1 = 1 - \frac{a_2 + a_3 + a_4}{r_E - 1} . \qquad (9b)$$

Thus, according to eqs. (8) and (9), the calculated value of k_1 is equal to 1.503 for PP-PC blends.

2. Polycarbonate – polymethylmethacrylate (PC-PMMA) blends.

The experimental data for PC-PMMA blends, taken from the literature [9,10], are shown in Table 3, as are the calculated values of volume fractions v and reinforcement factors e. The plot of e vs. v is shown in Fig. 3 (evidently outstanding points are omitted). Thus three types of blend, A,B and C, prepared and measured under different conditions, can be compared. We have assumed, in the first approximation, that the differences between the types are in the limits of experimental error. Hance the parallel (additive) model can be applied and the least-squares method gives e=0.192 v, with coefficient of multiple correlation being equal to 0.94 (cf. Table 4). However, 2-nd and 3-rd order equations for the suspension model can be assumed as well, with the same correlation, but with k_1=0.783 and k_1=1.146 respectively. For the 4-th order equation, the coefficient of multiple correlation was 0.35 and k_1=1.596.

The question which equation is the best will be discussed below.

Table 3. Data for PC-PMMA blends

| Experimental data | | Concentration of PMMA wt.% | Modulus $E\times10^{-3}$ MPa | Calculations | | | |
Method of preparation	Tested specimens			Density g/cm³	Volume fraction of PMMA v	e	ξ_R
A. Mechanical mixing of powdered components [9].	Injection moulded dumb-bell shaped specimens; thickness: 4 mm	0	2.25	1.20 (a)	0	0	-
		25	2.68	1.197	0.252	0.0510	0.9764
		40	2.94	1.196 (a)	0.402	|0.1529| (c)	-
		50	2.98	1.195	0.502	|0.1686| (c)	-
		60	2.85	1.194	0.602	0.1176	0.9707
		75	2.87	1.192 (a)	0.752	0.1255	0.9476
		100	3.14	1.19	1	0.2314	1
B. Mixing of solutions and coprecipitation [10].	Pressed sheets; thickness: 0.5 mm	0	1.68	1.20 (b)	0	0	-
		20	1.75	1.194	0.204	0.0417	1.0323
		40	1.70	1.188	0.406	|0.0119| (c)	-
		60	1.86	1.182	0.606	0.1071	1.0086
		80	1.94	1.176 (b)	0.804	0.1548	1.0221
		100	2.28	1.17 (b)	1	|0.3571| (c)	1
C. Polymerization of methyl methacrylate in solution of PC [10].	Pressed sheets; thickness: 0.5 mm	0	1.68	1.20 (b)	0	0	-
		20	1.76	1.194	0.204	0.0476	1.0571
		40	1.79	1.188	0.406	0.0655	0.9954
		60	1.88	1.182 (b)	0.606	0.1190	1.0254
		100	1.96	1.17 (b)	1	0.1667	1

(a) measured density; (b) taken from the literature [11]; (c) outstanding results.

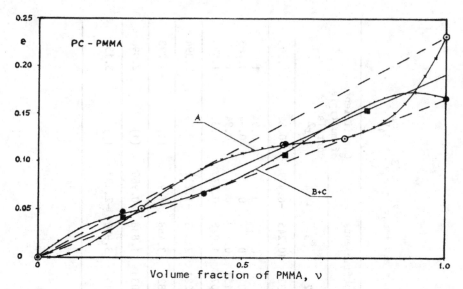

Figure 3. Plot of e vs. v for PC-PMMA blends.
Experimental data: (A) ⊙ mixed polymers [9];
(B) ■ mixed solutions, (C) ● PMMA polymerised in PC
solution [10]; ——— the parallel model for (A+B+C);
the 4th order equation for the suspension model:
(A) ✗✗✗ and (B+C) ••••.

DISCUSSION

It is evident from the examples described above that a 4-th
order equation for the suspension model fits the experimental data
for PP-PC blends best. In the case of PC-PMMA blends, however,
several equations of different order can be consistent with the
experimental data, and the choice of the best one is more difficult.
Coefficients of multiple correlation seem to be a rather inadequate
criterion here.

We have therefore attempted to compare the obtained results with
the parallel voids model (cf. Table 1), for which the reinforcement
function e can be written as

$$e = (\xi_R r_E - 1)v. \qquad (10)$$

Hance the fraction of the reinforcing material in the filler space
ξ_R is equal to

Table 4. Coefficients of equation $e = a_1v + a_2v^2 + a_3v^3 + a_4v^4$.

Polymer system	Set of data	Number of points	Order of equation	Coefficients				Coefficient of multiple correlation r^2	k_1
				a_1	a_2	a_3	a_4		
PP-PC	-	6	4	1.667	-4.324	10.245	-6.479	0.89	1.503
PC-PMMA	A+B+C	11	1	0.192	0	0	0	0.94	1
			2	0.148	0.041	0	0	0.94	0.783
			3	0.227	-0.152	0.123	0	0.94	1.146
			4	0.316	-0.676	1.028	-0.470	0.35	1.596
	B+C	7	4	0.395	-1.281	2.273	-1.222	0.79	2.394
	A	4	4	-0.068	1.826	-3.488	1.962	(1)	-0.293
	B	4	4	0.347	-1.033	1.838	-0.986	(1)	2.090
	C	4	4	0.526	-2.156	3.990	-2.193	(1)	3.150

$$\xi_R = 1/r_E + 1/r_E(e/v).$$ (11)

Eq. (11) combined with eq. (6) gives

$$\xi_R = b_o + b_1 v + b_2 v^2 + b_3 v^3 + \dots,$$ (12)

where

$$b_o = (a_1 + 1)/r_E ,$$ (12a)

and

$$b_i = \frac{a_i + 1}{r_E}$$ (12b)

The coefficients of eq. (12) can be calculated from the experimental data by the least-squares method.

It has recently been found [3] that k_1 in eq. (7) is approximately equal to ξ_R for v=0 in the system of glass-fiber filled PC. For polymer blends, taking $b_o = (\xi_R)_o$ as the value of ξ_R at v=0, and combining eqs. (9a) and (12a), we obtain

$$k_1 = \left[(\xi_R)_o \, r_E - 1\right]/(r_E - 1).$$ (13)

Then, substituting eq. (13) into eq. (7) and taking eqs. (12a) and (12b) into account, we obtain

$$e = \left[(\xi_R)_o \, r_E - 1\right]v + b_1 r_E v^2 + b_2 r_E v^3 + b_3 r_E v^4.$$ (14)

Eq. (14) is an extension of the voids model, given by eq. (10), and is equivalent to eqs. (6) or (7).

In our examples, the values of ξ_R were calculated from the experimental data shown in Tables 2 and 3, using eq. (11), and the results are shown in Figs. 4 and 5. The coefficients of eq. (12) were then calculated by the least-squares method and they are shown in Table 5. Thus, each system can be characterized by a function e=f(v), eqs. (6) or (14) (cf. Table 4), and by a corresponding function ξ_R=f(v), eq. (12) (cf. Table 5).

For the parallel model, we should have ξ_R=1 independently of v. This, however, is not observed for the PP-PC and PC-PMMA blends; therefore this model should be rejected for both systems. For PP-PC blends, the 4-th order equation for e=f(v) is confirmed by the 3-rd order equation for ξ_R=f(v). For PC-PMMA blends, group A and group B+C form two different populations, and 4-th order equations for e=f(v) or 3-rd order equations for ξ_R=f(v) also fit best the

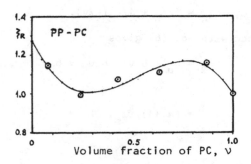

Figure 4.   Plot of ξ_R vs. v for PP-PC blends.

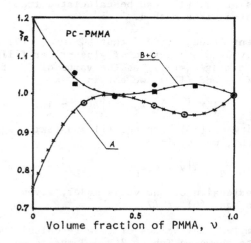

Figure 5.   Plot of ξ_R vs. v for PC-PMMA blends.
            Notation as in Fig. 3.

experimental data (cf. Tables 4 and 5 and Fig. 5). It is evident
that ξ_R is below unity for group A and is higher than unity for
group B+C (Fig. 5). The optimum properties for group A were found
for v ≈ 0.4 [9], which coincides with the maximum value of ξ_R.

According to the parallel voids model (Table 1) we have

$$\xi_R + \xi_L = 1,\qquad\qquad(15)$$

Table 5. Coefficients of 3rd order equation $\xi_R = b_o + b_1 v + b_2 v^2 + b_3 v^3$

Polymer system	Set of data	Number of points	$b_o = \xi_{R_o}$	b_1	b_2	b_3	Coefficient of multiple correlation r^2
PP–PC	–	6	1.272	-2.062	4.886	-3.090	0.89
PC–PMMA	A + B + C	11	1.126	-0.652	0.988	-4.633	0.24
	B + C	7	1.196	-1.098	1.947	-1.047	0.79
	A	4	0.757	1.485	-2.837	1.595	1
	B	4	1.154	-0.884	1.573	-0.843	1
	C	4	1.307	-1.846	3.415	-1.877	1

where ξ_R is the volume fraction of the reinforcing material in the
filler space, and ξ_L is the volume fraction of voids in the same
space. Thus the volume of reinforcing material is equal to

$$v_R = \xi_R v. \tag{16}$$

If $\xi_R=1$, voids are not present, and the matrix material perfectly
adheres to the surface of the reinforcing particles (Fig. 6b). If
$\xi_R<1$, we have voids in the filler space and the surface of reinforcing
particles is not completely coupled with the matrix material (Fig. 6a).
This is usually observed for polymers filled with hard particles of
another polymer or glass [1,3]. In this work we observed $\xi_R>1$. This
means that the reinforcing material occupies more space than its
original volume should do. Such a situation is possible for polymer
blends where macromolecules of both incompatible components are
penetrating their neighbouring material at the phase boundary. This
interpenetrated zone of mutual entanglements forms an interphase of
additional reinforcing material (Fig. 6c).

Therefore the fraction ξ_R can be considered as a measure of the
efficiency of the reinforcing material for a given two-component
polymer system, indicating also the state of adhesion at the phase
boundary. The efficiency of the blending method can thus be estimated, and the respective blends can be compared. For instance, for
the group A of PC-PMMA blends, about 3-5 vol.% of voids in the filler
space is observed, while in the case of B or C there is an excess of
filling volume, about 3 or 5 vol.% respectively (Fig. 5). The
efficiency of blend preparation increases in the sequence A-B-C, in
order of the increasing values of $(\xi_R)_0$: 0.757, 1.154 and 1.307
respectively (Table 5).

Moreover, for the system with $\xi_R>1$, the average interphase
thickness δ can be calculated from the following equation:

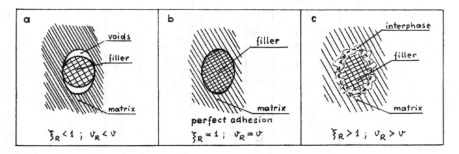

Figure 6. Interpretation of the fraction of reinforcing
material in the filler space, ξ_R.

$$\delta = \frac{d}{2}\sqrt[3]{\xi_R} - 1 \, , \qquad (17)$$

where d is the diameter of the spherical particles of the dispersed component. Eq. (17) is also approximately valid for rod-like dispersions of diameter d and length l=d(s+1) (see Fig. 7). Thus maximum observed values of δ are 0.008d and 0.024d for PC-PMMA and PP-PC blends, respectively.

For PP-PC blends, rod-like cylindrical PC dispersions with diameters of about 2-12 μm were observed by electron scanning microscopy [12]. Hence, the maximum average interphase thickness δ should be equal to about 0.05-0.30 μm.

CONCLUSIONS

The proposed reinforcement function for elastic modulus E of two-component polymer blends e=f(v,r_E, ...), where e=E/E_2-1, r_E=E_1/E_2 and v is the volume fraction of reinforcing component, can be conveniently used to compare different models. Experimental data obtained under different conditions can be compared since e is obtained as a relative value.

The suspension model can be expressed as a function:

$$e = a_1 v + a_2 v^2 + a_3 v^3 + a_4 v^4$$

and/or as a function

$$\xi_R = b_o + b_1 v + b_2 v^2 + b_3 v^3,$$

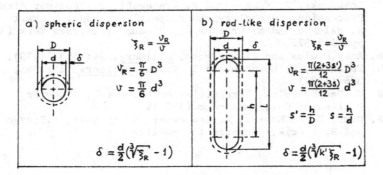

Figure 7. Calculation of interphase thickness, δ.

where ξ_R is the volume fraction of the reinforcing material in the
filler space. The numerical values of coefficients a and b can be
evaluated from experiments by measuring the moduli of pure components and their blends.

The results of statistical treatment of experimental data,
suggesting the additivity of the properties with high coefficient
of correlation, can conceal the true state of interactions between
components at the phase boundary. The function $\xi_R = f(v)$ can then be
convenient for a more detailed characterization of the system.

The fraction of reinforcing material in the filler space, ξ_R,
is the parameter of reinforcing efficiency and can be easily calculated from the experimental data. The efficiency of the blend preparation method can thus be estimated by means of ξ_R values or by
means of the limit value $(\xi_R)_o = b_o$. In the case of $\xi_R > 1$, the average
interphase thickness can be calculated.

It is possible to apply a similar procedure to investigate
other properties of polymer blends, using a reinforcement function
defined as $e = P/P_1 - 1$, where P is a given polymer property.

REFERENCES

1. J. A. Manson and L. H. Sperling,"Polymer Blends and Composites",
    Plenum Press, New York (1976).
2. J. Magryta and R. Gaczyński, _Polimery_, 23: 347 (1978).
3. Z. Dobkowski, 60th Meeting of Polish Chem.Soc., Wrocław, (1979);
    Z. Dobkowski, _Polimery_, 26: 10 (1981).
4. E. H. Kerner, _Proc.Phys.Soc._ 69B: 808 (1956).
5. J. C. Halpin and J. L. Kardos, _J.Appl.Phys._ 43: 2235 (1972).
6. E. Guth and O. Gold, Phys.Rev. 53: 322 (1938); E. Guth,
    J.Appl.Phys. 16: 20 (1945).
7. Z. Dobkowski, _Polimery_, 25: 110 (1980).
8. Z. Dobkowski, Z. Kohman and B. Krajewski, _in_ "Polymer Blends.
    Processing, Morphology and Properties", eds., E. Martuscelli,
    R. Palumbo and M. Kryszewski, Plenum Press, New York (1980),
    pp. 363-372.
9. H. Antychowicz and Z. Wielgosz, _Polimery_, 24: 193 (1979).
10. H. Polańska, T. Koomoto and T. Kawai, _Polimery_, 25: 365,
    396 (1980).
11. D. W. Van Krevelen, "Properties of Polymers", Elsevier
    Publishing Co., Amsterdam (1976).
12. A. Dems, Z. Dobkowski, B. Krajewski, J. Mejsner, W.Przygocki
    and H. Szocik, prepared for publication.

COMPATIBILITY OF PVC-POLYMETHACRYLATE SYSTEMS

M. Kozłowski and T. Skowroński

Technical University of Wrocław

Poland

Compatibility of two- and threecomponent systems of poly(vinyl chloride) with homo- and copolymers of methyl methacrylate and butyl methacrylate has been studied. The mixtures were tested for compatibility in the dynamic viscosity of polymer solutions, in microscopic observations in polarized light, as well as by dynamic mechanical measurements. It was found that poly(methyl methacrylate) and methacrylic copolymers were compatible with PVC. Poly(butyl methacrylate) appears to be incompatible with PVC. Estimation of solubility parameter values δ made it possible to predict the compatibility of polymer pairs. Critical $\Delta\delta$ value for compatible polymers has been found to be 0.5 (cal cm^{-3})$^{1/2}$.

INTRODUCTION

The range of application of traditional plastics can be considerably broadened by their physical modification, that is, by blending the polymer being modified with other low- or high-molecular weight compounds. Such multicomponent blends often possess interesting properties which is the reason for the observed increasing interest in their investigation and application.

The industrial practice of introducing new polymer compositions, which is dictated by market requirements, initially overtook the state of knowledge concerning multicomponent polymer blends, but the intensive investigations carried out during the past twenty years led to the formulation of some definitions and scientific

concepts[1-12]. Thus, the problem in question still remains a topical
one, and it has stimulated more optimal applications of polymer
blends.

The magnitude of intercomponent interactions is of basic importance in these systems, since it determines the structure of the
composition. Thus, in majority of works on the subject the problem
of the compatibility of polymers is discussed. Thermodynamic considerations indicate that compatibility is only seldom encountered in
polymer systems[1-12]. Free energy change ΔG_M, related to the formation
of a two-component polymer system under constant pressure and temperature conditions, is given by

$$\Delta G_M = \Delta H_M - T\Delta S_M \, , \tag{1}$$

where ΔH_M is the enthalpy of mixing, ΔS_M - entropy of mixing, and
T - absolute temperature of the system.

Two polymers are compatible in a solution only when $\Delta G_M < 0$;
in the case of $\Delta G_M > 0$ a two-phase system is formed. Entropy of
mixing of two polymers is very small, which follows from the
Flory-Huggins equation:

$$\Delta S_M = -k(n_1 \ln\phi_1 + n_2 \ln\phi_2) \, , \tag{2}$$

where n_1, n_2 is the number of polymer molecules, and ϕ_1, ϕ_2 - volume
fraction of polymers in the system. The larger the molecular weight
of a polymer pair, the lower their number in the unit volume and,
consequently, the lower their entropy of mixing. In view of the
fact that the term $T\Delta S_M$ always has positive values (molecular
association excluded), in order to obtain compatibility of two
polymers in a solution their heat of mixing has to have negative
values, be zero, or have a very small positive value ($<T\Delta S_M$).

For a system of two non-polar amorphous polymers, if the small
contraction of volume is omitted, the expression for the heat of
mixing has the following form:

$$\Delta H_M = \phi_1\phi_2 (\delta_1 - \delta_2)^2 \tag{3}$$

where δ_1, δ_2 are solubility parameters of the polymers.

From Eq. (3) it follows that $\Delta H_M > 0$, which indicates that two
polymers can mix only in the case when the values of their solubility parameters are very similar.

A number of experimental methods have been developed to determine polymer compatibility[13-19] e.g., observation of phase separation
in polymer solutions, mixed polymer solutions viscosity, and light

scattering measurements, glass transition temperatures, cast film
transparency, nonradiative energy transfer, permeation of vapors and
gases, various rentgenographic techniques, infrared spectroscopy, as
well as determination of rheological properties. Particularly
suitable are various microscopic techniques, i.e., optical and
electron microscopy which provides direct evidence about the
structure of multicomponent systems.

This paper considers the compatibility of poly(vinyl chloride)
(PVC) with homo- and copolymers of methyl methacrylate (MM) and
butyl methacrylate (BM), as well as the compatibility with blends
of homopolymers having compositions corresponding to the total
compositions of the copolymers.

EXPERIMENTAL

Materials

Suspension grade poly(vinyl chloride) Tarwinyl S-60, having the
Fikentscher number K = 60.3 and M_w = 1.02x10^5, as well as polymethacrylates prepared in our laboratory by suspension polymerization were
used throughout this study. The polymethacrylates used were PMMA,
PBMA and copolymers MMA/BMA; the respective comonomer ratios were
50/50 and 30/70. Molecular weights of all polymethacrylates were
similar (M_η = 9x10^4).

Preparation of Polymer Mixtures

The mixtures to be studied on the rotational viscometer were
prepared by dissolving varying amounts of the polymethacrylates and
PVC in dimethyl formamide at 10% by weight concentration. The
samples to be studied by microscopy were prepared by solvent
casting from the above mentioned solutions. The blends of PVC and
polymethacrylates to be tested on the Rheovibron instrument were
obtained by mechanical mixing of the components on a rolling mill,
followed by pressing to a foil form.

Measurements

The compatibility of the polymers in solutions was determined
by dynamic viscosity measurements using a rotational viscometer
at 293 K and shear rate 1312 s^{-1}. The morphology of the film
cast from a solution was examined with a polarization-interference
microscope by differential method, at magnification of 250x.
Dynamic mechanical measurements were made using direct reading
visco-elastometer, the Rheovibron. The temperature was scanned

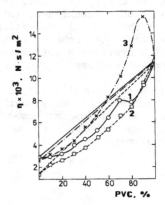

Fig. 1.   Dynamic viscosity of 10 wt.% blend solution vs. composition
          of PVC-polymethacrylate blends: 1-PMMA;   2-PBMA;
          3-MMA/BMA 50/50.

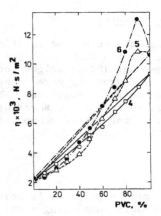

Fig. 2.   Dynamic viscosity of 10 wt.% blend solution vs. composition
          of PVC-polymethacrylate blends: 4-MMA/BMA 30/70;
          5-(PMM+PBMA)   50/50;   6-(PMMA+PBMA) 30/70.

from 193 K to 423 K (in the case of PMMA), the frequency being set at 110 Hz. The values of the solubility parameters of the polymers studied were determined by the viscometric method according to Gee[20] at 298 K, using solvents whose solubility parameters varied from 8.6 to 11.7 $(cal\ cm^{-3})^{1/2}$.

RESULTS AND DISCUSSION

The relationships between the dynamic viscosity of the solutions and the blend compositions are shown in Figs. 1 and 2. In order to interpret the obtained results the Zelinger-Heidingsfeld criterion[21,22] was used. Thus, only blends fulfilling the following condition were considered to be compatible:

$$\left| \frac{\Delta_i}{\Delta\eta_i} \right| < 0.1 \ , \tag{4}$$

where

$$\frac{\Delta_i}{\Delta\eta_i} = \frac{\eta_{ad} - \eta_{ex}}{\eta_A - \eta_B} \ , \tag{5}$$

where η_{ad} is additive viscosity, η_{ex} – experimental viscosity, and η_A, η_B – viscosity of the solution of polymer A and B, respectively. According to the criterion assumed, two-component blends of PVC with PMMA and PBMA are compatible in the range up to 10 wt.% and above 80 wt.% of polymethacrylates. PVC-MMA/BMA 30/70 (copolymer) system is compatible in the entire range of compositions, while PVC-MMA/BMA 50/50 only above 30 wt.% of the copolymer in the composition. Moreover, the three-component systems of PVC with methacrylic ester homopolymers are compatible within various composition ranges, e.g. when the content of polymethacrylates is higher than 80 wt.%.

Microscopic examination in polarized light of the foils cast from solutions seem to indicate that all compositions containing up to 30 wt.% of methacrylic polymers are compatible (the films are homogeneous). However, with their increasing content in the blends, an inhomogeneous structure becomes apparent. This was particularly evident for the PVC-PBMA and threecomponent blends of PVC with methacrylic homopolymers (Fig. 3). As PMMA content in the PVC-PMMA blends increases up to ca. 50-60 wt.%, the inhomogeneities of the system become most pronounced. Further increase in PMMA content makes the micrographs more uniform. Micrographs of the blends of PVC and MMA/BMA copolymers, especially with MMA/BMA 30/70 copolymer (Fig. 4), exhibit no inhomogeneities, suggesting compatibility of the system.

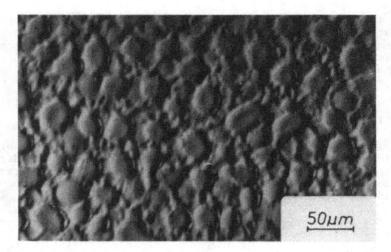

Fig. 3.   Microphotograph of 30/70 PVC-(PMMA+PBMA)   30/70 system.

        Samples of polymethacrylates mixed with PVC in ratios of 100:0,
5:95, 10:90, 15:85, 0:100 were examined on the Rheovibron. The
relationships between the tangent of the dynamic mechanical loss
angle and the temperature are shown in Figs. 5 and 6. From these
data, the glass transition temperature regions were established for

Fig. 4.   Microphotograph of 30/70 PVC–MMA/BMA 30/70 system.

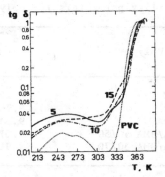

Fig. 5.   Temperature dependence of angle of dynamic-mechanical loss
          for PVC-(PMMA+PBMA) 30/70 system.

the blends studies by assuming that only one value of the glass
transition temperature is observed for a compatible system, and more
such T_g values exist in incompatible systems. In the light of this
criterion the blends of PVC with methacrylic copolymers as well as
the PVC-PMMA blend should be considered compatible. For PVC-PBMA
blends containing from 5 to 15 wt.% of PBMA, aside from a maximum
at 367 K corresponding to T_g of PVC, an inflection point on the curve
of dumping within a temperature range of 323-348 K was observed,
which can be related to T_g of PBMA. This indicates that the blend
is incompatible. A similar character of the dumping curves was also
observed for threecomponent systems composed of PVC and methacrylic
homopolymers. In this case the maximum on the dumping curve was
shifted toward higher temperatures with an increasing content of PMMA

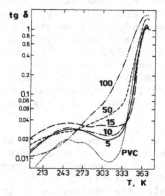

Fig. 6.   Angle of dynamic-mechanical loss vs. temperature for
          PVC-MMA/BMA 30/70 system.

Table 1.   Solubility parameters for polymers studied

Polymer	δ $(cal\ cm^{-3})^{1/2}$
PVC	10.22
PMMA	9.76
PBMA	9.51
MMA/BMA 50/50	9.87
MMA/BMA 30/70	9.98

in the blend. Such a relationship indicates that the systems are incompatible and that they are composed of two phases: one of pure PBMA, and one consisting of the blend of PMMA and PVC.

The results of compatibility studies presented above were confronted with the values of the solubility parameters for the polymers studied (Table 1). The differences between the solubility parameters for the copolymers studied and PMMA in relation to PVC were smaller than 0.5, while in the case of PBMA the difference was higher than 0.7. Yet again, it has been shown that on the basis of the values of the solubility parameters one can tell whether compatibility in a polymer system can be expected or not.

CONCLUSIONS

It has been found that MMA/BMA copolymers and poly(methyl methacrylate) are compatible with poly(vinyl chloride). Poly(butyl methacrylate) on the other hand, is incompatible with PVC. Threecomponent systems exhibit heterogeneous structures: it seems that one phase consisted of the blend of PMMA with PVC, while the other was PBMA. Differences have been observed in the results of compatibility tests between the mixtures in the solid state and in solution. A solvent, an additional component, may bring about drastic changes in the behaviour of the system, as changes of interactions between polymers may then occur.

A comparison of the results of compatibility studies with solubility parameters indicates that one may expect compatibility also in the case of polar polymers. In accordance with Eq. (3), the boundary value of compatibility $(\delta_1 - \delta_2)$ is 0.5 $(cal\ cm^{-3})^{1/2}$. Polymer pairs, for which the difference in the solubility parameters exceeds this value, should be incompatible.

REFERENCES

1. G. L. Slonimski, J.Polymer Sci., 30:625 (1958).
2. L. Bohn, Kolloid-Z., 213:55 (1966).
3. K. Friese, Plaste u. Kautsch., 12:90 (1965); 13:65 (1966);
   15:646 (1968).
4. J. A. Manson, L.H. Sperling "Polymer Blends and Composites",
   Plenum, New York (1976).
5. S. Krause, J. Macromol.Sci., C7:251 (1972).
6. M. Shen, H. Kawai, AIChE J., 24:1 (1978).
7. D. R. Paul, S. Newman "Polymer Blends", Academic Press,
   New York (1978).
8. R. Koningsveld, R.F.T. Stepto, Macromol., 10:1166 (1977).
9. L. P. Mc Master, ibid., 6:760 (1973).
10. D. Patterson, A. Robard, ibid., 11:690 (1978).
11. P. J. Flory "Principles of Polymer Chemistry", Cornell Univer-
    sity Press, Ithaca (1953).
12. R. L. Scott, J.Chem.Phys., 17:279 (1949).
13. Adv.Chem.Ser., N° 99 (1972).
14. B. D. Gesner, Appl.Polymer Symp., 7:53 (1968).
15. R. J. Kern, J.Polymer Sci., 21:19 (1956).
16. H. Morawetz, F. Amrani, Macromol., 11:281 (1978).
17. J. Stoelting, F.E. Karasz, W.J. Mc Knight, Polym.Eng.Sci.,
    10:133 (1970).
18. K. Kato, Polymer Letters, 4:35 (1966).
19. T. Hashimoto, K. Nagatschi, A. Todo, H. Hasegawa, H. Kawai,
    Macromol., 7:364 (1974).
20. G. Gee, Trans.Faraday Soc., 38:418 (1942).
21. J. Zelinger, V. Heidingsfeld, Sb.Vys.Sk.Chem.Technol.Praze
    Org.Technol., 9:63 (1966).
22. M. Rusu, D. Feldmann, Mater.Plastice, 12:146 (1975).

LINEAR POLYETHYLENE/POLYPROPYLENE/ETHYLENE-PROPYLENE COPOLYMER
TERNARY BLENDS:
I. RELATIONSHIP BETWEEN IMPACT PROPERTIES AND MORPHOLOGY

L. D'Orazio, R. Greco, C. Mancarella, E. Martuscelli,
G. Ragosta and C. Silvestre

Istituto di Ricerche su Tecnologia dei Polimeri e Reologia
del C.N.R., Arco Felice, Napoli

INTRODUCTION

It is well known that the mixing of two polymers leads generally
to heterogeneous systems since the compatibility between them is
rather a rare event[1-5]. Therefore multicomponent polymeric materials
result to be generally multiphase and obtainable in a variety of
morphologies depending on chemical and/or physical treatments. This
can be achieved in fact in some cases by adding to the binary blend
a third component such as a suitable graft or block copolymer or
a random copolymer[6-9]. In the present work the effect of the addition
of two commercial ethylene-propylene random copolymers (Dutral and
Epcar) with diverse ethylene percentage to binary high density polyethylene (HDPE)/isotactic polypropylene (IPP) blends has been investigated. Impact properties at room temperature and the overall
morphology of compression-molded samples have been studied. Also
a fractographic analysis by direct ispection of the specimens and
by a careful study of the morphology and structure of the fractured
surfaces has been effected by means of a scanning electron microscope.
The main goal of this paper is to elucidate the influence of composition and nature of added rubber on the rupture mechanisms as well
as on the overall morphology and on the interactions arising among
the components.

EXPERIMENTAL

The molecular characteristics of the polymers used for the
present study and the blend compositions explored are collected in
Tables 1 and 2 respectively. Details on the processing conditions
and on the specimens preparation are given in a previous paper[10].

Table 1. Characteristic of polymers used

Code	$\overline{M}_w$	$\overline{M}_w/\overline{M}_n$	MFI g/10 min.	C_2 % in moles	Tg °C	Source
HDPE	1.1×10^5	10	3.7	–	$-90 \div -140$	Rapra
iPP	3.1×10^5	20	3.9	–	$-10 \div 0$	Rapra
Dutral	1.8×10^5	–	–	60	-60	Montedison
Epcar	1.4×10^5	3.4	–	74	-47	Goodrich

Table 2. Blend Composition

Blend code HDPE/iPP/copolymer	HDPE (%)	iPP (%)	copolymer (%)	$Y_{PE} = \dfrac{W_{HDPE}}{W_{HDPE} + W_{iPP}}$
100/0/0	100	–	–	1
95/0/5	95	–	5	–
90/0/10	90	–	10	–
85/0/15	85	–	15	–
75/25/0	75	25	–	0.75
71.25/23.75/5	71.25	23.75	5	–
67.5/22.5/10	67.5	22.5	10	–
63.75/21.25/15	63.75	21.25	15	–
50/50/0	50	50	–	0.50
47.5/47.5/5	47.5	47.5	5	–
45/45/10	45	45	10	–
42.5/42.5/15	42.5	42.5	15	–
25/75/0	25	75	0	0.25
23.75/71.25/5	23.75	71.25	5	–
22.5/67.5/10	22.5	67.5	10	–
21.25/63.75/15	21.25	63.75	15	–
0/95/5	–	95	5	–
0/90/10	–	90	10	–
0/85/15	–	85	15	–
0/100/0	–	100	–	0

Izod impact tests at room temperature were performed on compression molded samples (50x13.0x3.0 mm^3), having a notch of 0.5 mm deep and a tip radius of 0.25 mm. The fractured surfaces of the broken specimens were examined in a scanning electron microscope (SEM) after coating with Au-Pd by means of an evaporator.

RESULTS AND DISCUSSION

Impact Properties

The Izod impact strength (R) of binary and ternary blends containing Epcar (solid lines) and Dutral (dotted lines) copolymers are shown in Fig. 1, as a function of the binary HDPE weight ratio defined as:

$$Y_{PE} = W_{HDPE}/(W_{HDPE} + W_{iPP}),$$

where W_{HDPE} and W_{iPP} are the weights of HDPE and iPP respectively. The copolymer percentage is indicated on the curves. The binary HDPE/iPP alloys show R values slightly lower than those of the two homopolymers. The binary HDPE/copolymer blends show a very drastic increase in the impact resistance with increasing copolymer content. Such an effect is even more marked in the case of Epcar mixtures, for which at concentrations higher than 10 percent no rupture is possible at the working conditions. A smaller increase in the impact strength

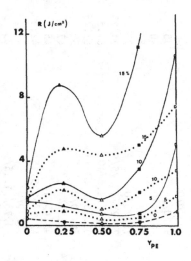

Fig. 1. Impact resistance R vs. Y_{PE} for Dutral (——) and Epcar (----) blends (copolymer percentage as indicated).

values with the increase of copolymer is observed in the case of iPP/copolymer binary blends. The shape of the curves corresponding to the ternary alloys shows the presence of a relative maximum at $Y_{PE} = 0.25$, a relative minimum at $Y_{PE} = 0.50$, and a monotonic increase of R at values higher than 0.50. In addition to this complex behaviour it must be noted that the ternary blends exhibit impact strength values higher than those of the corresponding binary HDPE/iPP blends and of the pure homopolymers. These features are generally more marked in the case of Epcar containing alloys and particularly in the HDPE matrix regions.

Fractographic Analysis: Fracture Mode and Mechanism

The negatives of the photographs of the ruptured specimens are shown in Figs. 2 and 3. The black zones around the tip of the notch represent in negative the stress whitening due to the formation of multicrazes during the impact test. For blends with high HDPE content it was possible by inspection to observe also a transverse contraction of the sample around the tip zone occuring before the fracture initiation. Further details of such figures will be discussed in the following paragraphs in comparison with SEM analysis features.

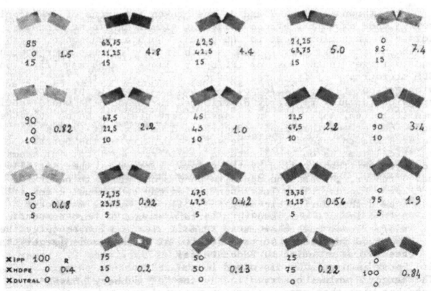

containing blends. Composition and R values as indicated.

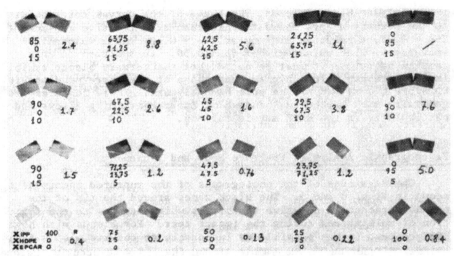

Fig. 3.   Negative of photographs of ruptured specimens of Epcar
          containing blends.   Composition and R values as indicated.

Binary iPP/copolymers blends

      As shown in Figs. 2 and 3, the broken specimens of iPP/copoly-
mers blends exhibit a stress whitening effect due to multicrazes
formation in regions near to the notch.   Such a phenomenon, which
is absent for pure iPP, increases with increasing copolymer content
and is larger in the case of iPP/Epcar alloys.   High magnification
SEM micrographs of fractured surfaces (relative to the fast crack
propagation area) of iPP/copolymers (85/15) show the presence of
spherical domains regularly distributed throughout the whole specimen (Figs. 4a, 4b).   As can be seen, there seems to be no evidence of
adhesion between the dispersed phase and matrix.   It is also
interesting to observe that the surface density of the Dutral
spherical domains on the fractured sample surface (Fig. 4a) seems
to be larger than implied by the volume fraction of the initially
added rubber, whereas the opposite phenomenon seems to occur for
iPP/Epcar blends.   This last observation may be accounted for if it
is assumed that during blending the following events may occur:
i)   Dutral is able to dissolve a certain amount of more defective
     and/or lower molecular mass iPP molecules and consequently the
     observed fractional surface density of particles is larger than
     the volume fraction of added Dutral.
ii) Epcar is partially dissolved in the iPP rubbery phase.
The larger R values observed in the case of iPP/Epcar blends,
together with the larger volume of material related to the stress
whitening phenomenon, may be attributed to the fact that in such

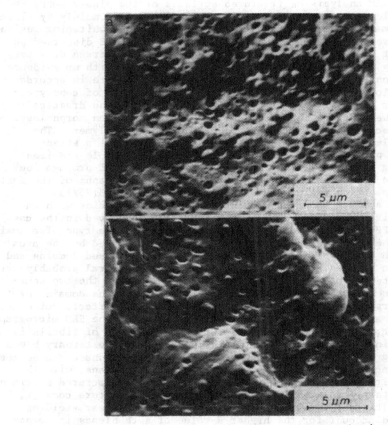

Fig. 4.   SEM micrographs of fractured surfaces of iPP/copolymer
          blends: a) iPP/Dutral (85/15) blend;   b) iPP/Epcar
          (85/15) blend.

blends the process of multicrazes formation and termination is more effective owing to the larger interparticle distances. The modification of the iPP rubbery regions (plastification) may also be responsible for the better impact behaviour of iPP/Epcar blends.

Binary HDPE/iPP (25/75) and ternary HDPE/iPP copolymer blends

The SEM analysis of fractured surfaces of the binary HDPE/iPP (25/75) blend shows that fracture propagation occurs mainly by direct crack formation without crazing. In fact no stress whitening regions are observed in Figs. 2 and 3. Furthermore, the HDPE dispersed phase is present as droplet; like domains (about 2-3 μm average diameter) uniformely distributed throughout the whole sample with no evidence of adhesion to the iPP matrix. These observations are in accordance with the low R values of such a blend. The addition of copolymers (Dutral or Epcar) to the binary HDPE/iPP (25/75) blend drastically changes the fracture mechanism and the overall surface morphology. These effects depend on the nature of the added copolymer. The ternary blend containing 15% of Dutral (Fig. 2) shows a stress whitening phenomenon that extends over nearly the whole specimen suggesting a very efficient multicraze production and propagation during the fracture. High magnification SEM micrographs of the fast crack propagation region of HDPE/iPP/Dutral (21.25/63.75/15) fractured surface show the presence of spherical domains with an average size (about 1-1.5 μm) lower than that observed in the case of HDPE/iPP binary blend (Fig. 5a). Evidence of some type of adhesion between the dispersed phase and the matrix is provided by the area micrograph (Fig. 5b) where fibrils connecting dispersed domains and matrix can be seen. These results indicate that Dutral probably acts as an "interfacial agent" favouring adhesion between the two semi-crystalline polymers and reducing the dimension of the domains of the dispersed phase. Owing, among others, to this effect, quick crack formation and propagation is avoided. As shown in the SEM micrograph of Fig. 5c, a clear induction area with the presence of fibrils is observed in the case of the fractured specimen of the ternary blend containing 15% of Epcar. Furthermore, no dispersed phase can be seen on the fractured surface (Fig. 5d). These observations, with the stress whitening phenomenon (Fig. 3) shown by the fractured specimens, indicate that for such blends two mechanisms of fracture coexist, namely multicraze formation (and propagation) and shear yielding. This could account for the higher R value of such blends in comparison with Dutral containing ternary blends.

Binary HDPE/iPP (50/50) and ternary HDPE/iPP/copolymer blends

The broken specimens of binary HDPE/iPP (50/50) blend do not exhibit the stress whitening phenomenon (see Figs. 2 and 3). As shown in the SEM micrograph of Fig. 6a HDPE and iPP form interconnected domains. Furthermore on the fractured surface it is possible to see fibrils, which seem to connect different domains,

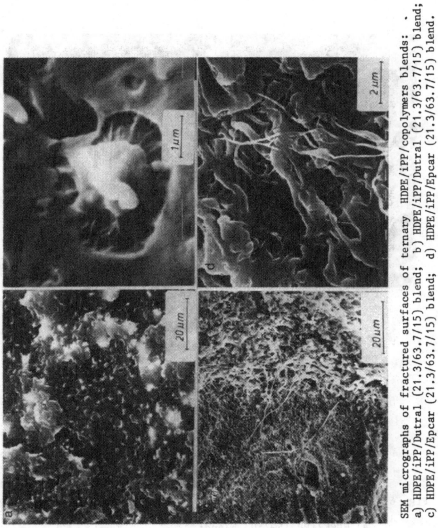

Fig. 5.   SEM micrographs of fractured surfaces of ternary HDPE/iPP/copolymers blends: .
a) HDPE/iPP/Dutral (21.3/63.7/15) blend;   b) HDPE/iPP/Dutral (21.3/63.7/15) blend;
c) HDPE/iPP/Epcar (21.3/63.7/15) blend;   d) HDPE/iPP/Epcar (21.3/63.7/15) blend.

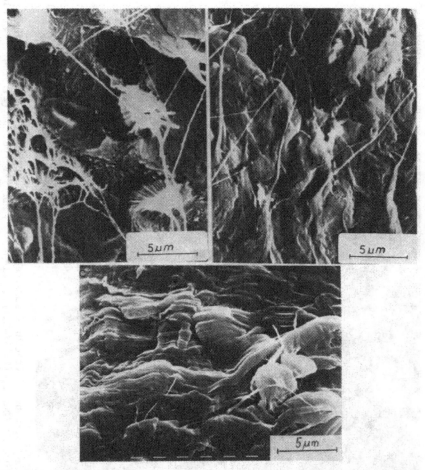

Fig. 6.   SEM micrographs of fractured surfaces of binary HDPE/iPP
          (50/50) and ternary HDPE/iPP/copolymer blends:
          a) HDPE/iPP (50/50) blend;
          b) HDPE/iPP/Dutral (42.5/42.5/15) blend;
          c) HDPE/iPP/Epcar (42.5/42.5/15) blend.

indicating that part of the material, presumably HDPE, has been
plastically deformed. The addition of copolymers (Dutral or Epcar)
modifies both the fracture mechanism and the overall surface
morphology. Fractured specimens of the ternary blends containing
Dutral show a stress whitening phenomenon that increases with
increasing copolymer content suggesting a multicraze fracture
mechanism (Fig. 2). As shown in Fig. 6b the macromorphology of the
fractured surface of ternary blends containing 15% of Dutral is more
regular in comparison with that observed in the case of HDPE/iPP
(50/50) binary blend. No distinction between the phases is observed
even though a fibrillation phenomenon is present, indicating that
a shear yielding mechanism is also active (Fig. 6b). Addition of
Epcar to the binary HDPE/iPP (50/50) blend produces no multicraze
effect (see Fig. 3) but fibrils and plastically deformed material
appear on the fractured surface of blend specimens. This strongly
suggests that shear yielding is the predominant fracture mechanism.
Finally, even for such blends no distinct phases are seen on the
broken surfaces (Fig. 6c). From the above it may be concluded that
the addition of rubbers to the binary HDPE/iPP (50/50) blends induces
a more ductile fracture behaviour. Which ductile fracture mechanism
that becomes activated depends on the chemical nature of the added
rubber.

Binary HDPE/iPP (75/25) and ternary HDPE/iPP/copolymer blends

The impact behaviour of HDPE/iPP (75/25) binary blend is of
brittle nature with no stress whitening phenomenon. As shown in the
SEM micrograph of Fig. 7a, the fractured surface of this binary
mixture exhibits a "honeycomb" morphology with evidence of dispersed
particles (iPP) enclosed in the cells. Furthermore, there seems to
be very negligible adhesion between the matrix and the dispersed
phase. This observation accounts for the lower impact strength
values compared with those obtained for the pure homopolymers.
Also in this case the addition of copolymer (Dutral or Epcar)
drastically changes the fracture mechanism and the morphology of
this system. The ternary blends containing Dutral show (Fig. 2)
a stress whitening phenomenon that increases with the increase of
the copolymer content indicating the formation of multicraze path-
ways during the fracture tests. As shown in Fig. 7b the whole
surface of HDPE/iPP/Dutral (63.75/21.25/15) fractured sample is
covered with fibrils suggesting that the sample undergoes plastic
deformation and that a ductile fracture mechanism is active (shear
yielding). In fact, for this specimen a phenomenon of lateral
contraction is also observed in Fig. 2. Furthermore, no evident
segregation phenomenon of the dispersed phase is observed (Fig. 7b).
For the ternary Epcar blends, the broken specimens (Fig. 3) present
a lateral contraction in regions close to the notch with a limited
material volume related to the stress whitening phenomenon. At room
temperature the ternary HDPE/iPP/Epcar (63.75/21.25/15) alloy does
not break completely during the Izod test. The SEM micrograph of

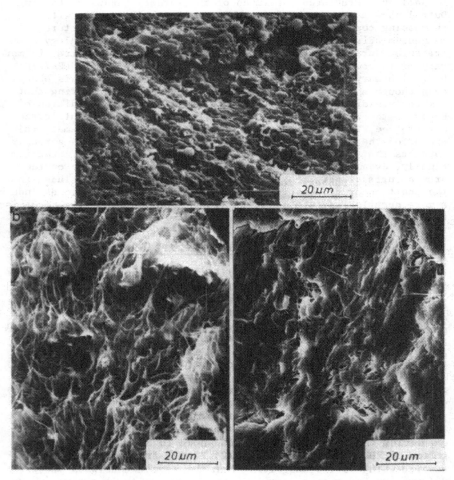

Fig. 7. SEM micrographs of fractured surfaces of binary HDPE/iPP
         (75/25) and ternary HDPE/iPP/copolymer blends:
         a) HDPE/iPP (75/25) blend;
         b) HDPE/iPP/Dutral (63.7/21.3/15) blend;
         c) HDPE/iPP/Epcar (63.7/21.3/15) blend.

such a ternary mixture shows extensive presence of fibrils on the
slow and fast crack propagation area Fig. 7c , indicating a ductile
fracture mechanism. It is to be emphasized that also for this
blends there is no evidence of the dispersed phase. All these
results suggest that the addition of a certain amount of Dutral
or Epcar   10% to the binary blend HDPE/iPP (75/25) produces the
following effects:

i)   homogeneization of the materials as no dispersed phase
      is observed;

ii)   increase of the impact resistance values as the materials
      become more ductile;

iii) a change of the fracture behaviour from a brittle crazing
      mechanism to a combination of shear yielding and multicraze.
      In other words both Dutral and Epcar seem to be able to act
      as "emulsifying agents" for the two semicrystalline polymers.

Binary HDPE/copolymer blends

     Broken specimens of pure HDPE show neither stress-whitening
effect nor macroscopic lateral contraction (Figs. 2 and 3). As
shown in the SEM micrograph of Fig. 8a, the fracture surface of
pure HDPE exhibits the presence of fibrils suggesting that the
sample undergoes also some plastic deformation. This observation
leads to the conclusion that HDPE moulded samples show features
of essentially brittle fracture mechanism with some possibility
of yielding. The fracture mechanism changes drastically when both
Dutral and Epcar are added. At room temperature HDPE/Epcar, 90/10,
85/15 and HDPE/Dutral 85/15 blends do not completely break and a
lateral contraction in regions close to the fractured surfaces can
be observed together with a slight stress whitening effect along
such surfaces (Figs. 2 and 3). SEM micrographs of the fractured
surface of HDPE/copolymers blends show a distinct induction area
characterized by a fibrillation phenomenon and a macromorphology
of the propagation crack region marked mainly by the presence of
tears. Furthermore, as shown in Figs. 8b and 8c no segregated
dispersed phase can be observed in such binary blends. These observations suggest that the copolymer chains are at least partially
dissolved in the HDPE amorphous regions. This conclusion would
indicate a certain "compatibility" between the HDPE amorphous phase
and the two EPM copolymers. The dissolution of copolymer chains
in HDPE amorphous phase would cause a drastic change in the matrix
properties. In fact, this effect can account for the higher resi-
lience values of HDPE/copolymers blends in comparison with those
observed for pure HDPE. Finally it is to be observed that binary
blends containing Epcar exhibit impact resistance higher than
those containing Dutral.

Fig. 8.   SEM micrographs of fractured surfaces of pure HDPE
          and HDPE/copolymer blends:
          a) HDPE;
          b) HDPE/Dutral (85/15) blend;
          c) HDPE/Epcar (85/15) blend.

CONCLUSIONS

The impact behaviour of compression moulded specimens of HDPE and iPP homopolymers and of HDPE/iPP binary blends may be strongly improved by the addition of the EPM copolymer. This is related to the fact that both the overall morphology and the mechanism and mode of fracture are deeply modified by the presence in the sample of a certain amount of rubber. Furthermore, it may be concluded that both the overall morphology and the fracture mechanism are strongly dependent on factors such as the nature of the matrix, (HDPE/or iPP), the composition and the chemical structure and the molecular mass of the added copolymer. It is interesting to point out that EPM copolymers seem to have the capability of acting as "interfacial agents" improving for example the adhesion between HDPE dispersed phase and iPP matrix. These observations suggest that a certain amount of EPM molecules can be "dissolved" in the amorphous phase of HDPE and iPP. Finally the finding that in binary HDPE/EPM copolymers blends no dispersed phase is observed, contrary to what we found in the case of iPP/EPM copolymers blend, suggests that the affinity of EPM molecules is likely to be higher with HDPE molecules than with iPP ones.

ACKNOWLEDGMENT

This research was supported by a grant from the Progetto Finalizzato Chimica Fine e Secondaria of the National Research Council (C.N.R.).

REFERENCES

1. D. R. Paul and S. Newman, "Polymer Blends", Academic Press, New York (1978).
2. D. Deanin, A. Deanin, T. Joblom, in: "Recent advances in polymer blends, grafts and blocks", L.H. Sperling, ed., Plenum Press, New York (1974).
3. J. A. Manson, L.H. Sperling, "Polymer Blends and Composites", Plenum Press, New York (1976).
4. R. Greco, G. Mucciariello, G. Ragosta, E. Martuscelli, J.Mat.Sci., 15:625 (1973).
5. Ibid., 16:1001 (1981).
6. R. E. Robertson, D.R. Paul. J.Appl.Polym.Sci., 17:2579 (1973).
7. D. R. Paul, "Interfacial Agent for copolymer blends" in: "Polymer Blends", D.R. Paul and S. Newman, eds., Academic Press, New York (1978).
8. E. Martuscelli, R. Palumbo and M. Kryszewski, "Polymer Blends: Processing, Morphology and Properties", Plenum Press, New York (1980).

9. E. Nolley, J.W. Barlow, D.R. Paul, Polym.Eng.Sci., 20:364
     (1980).
10. L. D'Orazio, R. Greco, C. Mancarella, E. Martuscelli,
     G. Ragosta, C. Silvestre, Polym.Eng.Sci., 22:536 (1982).

LINEAR POLYETHYLENE/POLYPROPYLENE/ETHYLENE-PROPYLENE COPOLYMER TERNARY BLENDS:
II. RELATIONSHIP BETWEEN MECHANICAL PROPERTIES AND MORPHOLOGY

L. D'Orazio, R. Greco, E. Martuscelli and G. Ragosta

Istituto di Ricerche su Tecnologia
dei Polimeri e Reologia del C.N.R.,
Via Toiano 6, 80072 ARCO FELICE, Napoli, Italy

INTRODUCTION

From a mechanical point of view, high density polyethylene (HDPE)/isotactic polypropylene (iPP) blend have generally been considered as very unsatisfactory materials[1,3]. In particular, they show very poor ultimate mechanical properties at room temperature in comparison with those of the blend constituents. This fact precludes their use for most commercial purposes. In previous papers[4,5] we found that it is possible to improve the mechanical tensile performance of these blends by appropriately varying the testing conditions such as the temperature and the drawing rate. In this way even ultradrawn blend fibers with high modulus and strength could be easily obtained, just as in the case of pure polyolefins.

The ultimate mechanical properties of HDPE/iPP blends can also be improved by adding a suitable third component, mainly a random ethylene-propylene copolymer (EPM). Such an additive, having intermediate molecular characteristics between the two polymer species, can act as a "compatibilizing agent" in their amorphous regions. The role of EPM copolymers as "compatibilizing agents" for polyolefin blends was first suggested by Paul et al.[7]. They showed, furthermore that copolymers with a certain degree of residual ethylene crystallinity were more effective than those having almost no crystallinity. We came to the same conclusions in a previous work in which we explored the effect of the addition of two of these copolymers (Dutral and Epcar) on the impact performance and on the morphology of HDPE/iPP blends[6]. The primary concern of this paper are the mechanical tensile properties at room temperature and the morphology of extruded samples of binary HDPE/iPP and ternary

127

HDPE/iPP/EPM blends. The main goal of the study was to clarify the influence of EPM copolymers on the morphology and on the mechanical and ultimate properties of extruded blends.

EXPERIMENTAL

     The polymers, whose characteristics are summarized in Table 1, were melt mixed in a Brabender-like apparatus at 200° C and at two residence times: 6 min. at 2 r.p.m. and further 10 min. at 32 r.p.m. The blend compositions are listed in Table 2. After premixing, cylindrical specimens were obtained directly by extrusion using a melting-elastic miniextruder (CSI max mixing extruder mod. CS-194). Thermal and tensile mechanical tests were performed on these specimens by an Instron Machine (mod. 1122) at room temperature and at cross-head speed of 10 mm/min. Also made were morphological studies by optical microscopy of sections microtomed from tensile samples and scanning electron microscopy of fractured surfaces of samples broken at liquid nitrogen temperature. Further details on the experimental procedures and on the techniques used are reported elsewhere[8].

RESULTS AND DISCUSSION

Thermal analysis

     Melting point (T_m) and fractional crystallinity (X_c) of HDPE and iPP for all the examined blends were detected by differential scanning calorimetry. Small variations of T_m and X_c were observed with increasing Dutral or Epcar copolymer content. Only for the HDPE/Epcar mixtures does the crystallinity of HDPE seem to increase with growing copolymer content. This effect may be attributed to the fact that Epcar has a high C_2/C_3 ratio. Therefore its longest ethylene sequences may crystallize contributing to some extent to the overall crystallinity.

Table 1. Characteristics of polymers used

Code	$\overline{M}_w$	$\overline{M}_w/\overline{M}_n$	MFI g/10 min.	C_2 % in moles	Tg $^\circ$C	Source
HDPE	1.1×10^5	10	3.7	–	$-90 \div -140$	Rapra
iPP	3.1×10^5	20	3.9	–	$-10 \div 0$	"
Dutral	1.8×10^5	–	–	60	-60	Montedison
Epcar	1.4×10^5	3.4	–	74	-47	Goodrich

Table 2. Blend composition

Blend code HDPE/iPP/copolymer	HDPE (%)	iPP (%)	Copolymer (%)	$Y_{PE} = \dfrac{W_{HDPE}}{W_{HDPE} + W_{iPP}}$
100/0/0	100	–	–	1
95/0/5	95	–	5	–
90/0/10	90	–	10	–
85/0/15	85	–	15	–
80/0/20	80	–	20	–
75/0/25	75	–	25	–
70/0/30	70	–	30	–
75/25/0	75	25	–	0.75
71.25/23.75/5	71.25	23.75	5	–
67.5/22.5/10	67.5	22.5	10	–
63.75/21.25/15	63.75	21.25	15	–
60/20/20	60	20	20	–
56.25/18.75/25	56.25	18.75	25	–
52.5/17.5/30	52.5	17.5	30	–
50/50/0	50	50	0	0.50
47.5/47.5/5	47.5	47.5	5	–
45/45/10	45	45	10	–
42.5/42.5/15	42.5	42.5	15	–
40/40/20	40	40	20	–
37.5/37.5/25	37.5	37.5	25	–
35/35/30	35	35	30	–
25/75/0	25	75	0	0.25
23.75/71.25/5	23.75	71.25	5	–
22.5/67.5/10	22.5	67.5	10	–
21.25/63.75/15	21.25	63.75	15	–
20/60/20	20	60	20	–
18.75/56.25/25	18.75	56.25	25	–
17.5/52.5/30	17.5	52.5	30	–
0/100/0	–	100	–	0
0/95/5	–	95	5	–
0/90/10	–	90	10	–
0/85/15	–	85	15	–
0/80/20	–	80	20	–
0/75/25	–	75	25	–
0/70/30	–	70	30	–

Tensile mechanical properties

Stress-strain curves for iPP/Dutral and iPP/Epcar blends are
shown in Figs. 1a and 1b respectively. In both cases a decrease in
the modulus and yield stress with increasing copolymer content is
observed. However, the influence of Epcar seems to be greater
since the yielding peak tends to disappear more rapidly than in the
blends containing Dutral. In Figs. 2 and 3 presented Young's moduli
for all the examined mixtures are as a function of copolymer content.
For Dutral containing blends, the modulus values (Fig. 3) decrease
gradually with an almost linear trend, whereas Epcar blends (Fig. 4)
show an initial sharp drop in the modulus as soon as a small amount
of copolymer (5%) is added, followed by a further slight decrease
when the copolymer content is raised. As could be expected, the
lowering of the modulus is followed by a decrease in the corresponding
tensile yield stress values (σ_y). In fact, as shown in Figs. 4 and 5,
both Dutral and Epcar blends exhibit an almost linear decrease of σ_y
with increasing copolymer concentration. The tensile strength (σ_r)
of binary and ternary blends containing Dutral and Epcar are shown
in Figs. 6 and 7 as a function of the binary HDPE weight ratio,
defined as $Y_{PE} = W_{HDPE}/(W_{iPP} + W_{HDPE})$ (where W_{HDPE} and W_{iPP} are the
weights of HDPE and iPP respectively in the blends). Binary HDPE/iPP
blends show a marked decrease in σ_r with a minimum at about $Y_{PE} =$
$= 0.75$. The addition of Dutral to pure HDPE (Fig. 6) causes a decrease in σ_r, whereas for iPP/Dutral mixtures the σ_r values are about
equal to those of the pure iPP. For the ternary blends containing
Dutral, generally an improvement in σ_r is observed with respect to
the binary HDPE/iPP. The extent of such an effect depends on the
Y_{PE} values as well as on copolymer content. For Epcar containing
blends as shown in Fig. 7, at $Y_{PE} = 0.50$ all the ternary mixtures
exhibit a pronounced maximum, whose value decreases with increasing
Epcar amount. A further feature to be underlined is the fact that
the addition of Epcar to both homopolymers produces a decrease in
σ_r values, whereas the combined effect of the HDPE/iPP ratio and
the copolymer give σ_r values very close to those of pure homopolymers. It is to be pointed out, however, that for binary HDPE/iPP
blends the rupture of the specimens occurred during the cold drawing
and it was caused by instability of the flow, probably due to the
interactions arising between the crystallities of the two polymers
and their connecting tie molecules. On the other hand, for homopolymers and blends containing the copolymer (Dutral or Epcar), the
tensile strenght values were related to fibre rupture. This last
finding may be due to the fact that the copolymer is able to reduce
the instability of flow making the cold drawing process easier and
more continuous, as in the case of pure homopolymers.

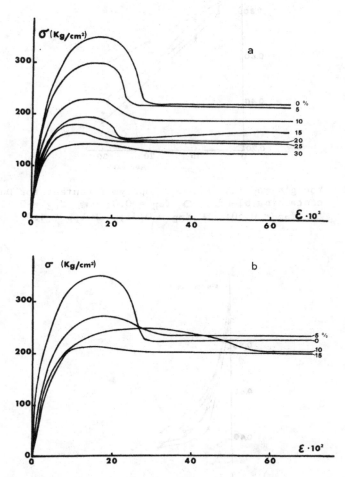

Fig. 1.   Stress-strain curves: a) iPP/Dutral blends;   b) iPP/Epcar blends.   Copolymer content as indicated.

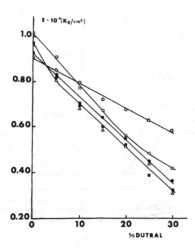

Fig. 2. Young's modulus (E), vs. copolymer content for Dutral
containing blends: ○ $Y_{PE} = 0.0$; ● $Y_{PE} = 0.25$;
$Y_{PE} = 0.50$; ■ $Y_{PE} = 0.75$; □ $Y_{PE} = 1.0$.

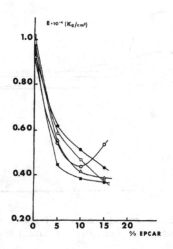

Fig. 3. Young's modulus (E), vs. copolymer content for Epcar
containing blends: ○ $Y_{PE} = 0.0$; ● $Y_{PE} = 0.25$;
△ $Y_{PE} = 0.50$; ■ $Y_{PE} = 0.75$; □ $Y_{PE} = 1.0$.

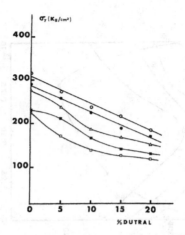

Fig. 4. Tensile yield stress (σ_y) vs. copolymer content for Dutral
containing blends: O Y_{PE} = 0.0; ● Y_{PE} = 0.25;
△ Y_{PE} = 0.50; ■ Y_{PE} = 0.75; □ Y_{PE} = 1.0.

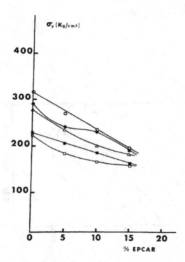

Fig. 5. Tensile yield stress (σ_y) vs. copolymer content for Epcar
containing blends: O Y_{PE} = 0.0; ● Y_{PE} = 0.25;
△ Y_{PE} = 0.50; ■ Y_{PE} = 0.75; □ Y_{PE} = 1.0.

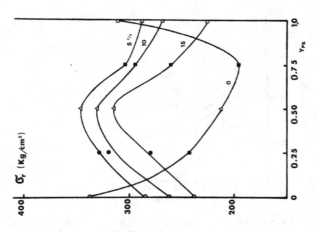

Fig. 7. Tensile strength (σ_r) as a function of
Y_{PE} for Epcar containing blends (copolymer content as indicated): $\bigcirc Y_{PE}$ =
= 0.0; $\bullet Y_{PE}$ = 0.25; $\triangle Y_{PE}$ = 0.50;
$\blacksquare Y_{PE}$ = 0.75; $\square Y_{PE}$ = 1.0.

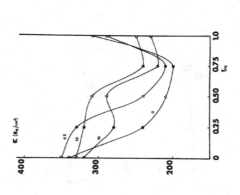

Fig. 6. Tensile strength (σ_r), as a function
of Y_{PE} for Dutral containing blends
(copolymer content as indicated):
$\bigcirc Y_{PE}$ = 0.0; $\bullet Y_{PE}$ = 0.25; $\triangle Y_{PE}$ =
= 0.50; $\blacksquare Y_{PE}$ = 0.75; $\square Y_{PE}$ = 1.0.

Morphological analysis of extruded samples

iPP/copolymers and HDPE/copolymers binary blends

Optical analysis of pure iPP and HDPE extrudates shows that
the two semicristalline polymers exhibit spherulitic and microspherulitic structure respectively (Figs. 8 and 9). It is to be noted
that the HDPE microspherulites are oriented along telescopic decen-
tralized patterns. Such circular orientations may be ascribed to
the sudden air quenching of the molten material at the die exit.
The addition of both copolymers to iPP produces a decrease in the
average size of the spherulites. This effect increases with
increasing copolymer content in the blend (Fig. 10). In agreement
with the results obtained by Karger et al.[9] and by Martuscelli
et al.[10], the above observation suggests that EPM copolymers may
act as nucleating agents for iPP spherulites.

Binary HDPE/iPP and ternary HDPE/iPP/copolymers blends

Optical micrographs of cross sections of the extruded HDPE/iPP
(25/75) filaments show a fine morphological structure with microspherulites aligned along telescopic rings (Fig. 11). In the decen-
tralized centre of the filament there is a zone of almost circular
shape with different morphology; this region contains mainly HDPE.
In fact, at temperatures higher than the T_m of HDPE but lower than
that of iPP, the material in this zone melts and birefringence
disappears. The segregation of the dispersed phase in the middle
of the filament is certainly due to the fact that the HDPE/iPP
(25/75) blend undergoes a particular (melt elastic) extrusion
process[11,12]. An analogous phenomenon was found by Martuscelli
et al.[13] while studying a HDPE/aPS (25/75) blend processed by the
same extruder. It is interesting to point out that by adding a small
amount (5%) of Dutral or Epcar both orientation and segregation
effects disappear. As shown in Fig. 12 a homogeneous distribution
of iPP and HDPE spherulites throughout the entire cross-section of
the filaments is obtained. Optical analysis of the extruded HDPE/iPP
(50/50) and HDPE/iPP (75/25) filaments shows that the structure of
the two components is microspherulitic (Fig. 13). Furthermore, no
orientation in the cross sections and no segregation effects are
observed. Typical scanning electron micrographs of fracture surfaces
of HDPE/iPP (25/75) and HDPE/iPP (75/25) extruded filaments broken
in liquid N_2 are presented in Figs. 14 and 15. As can be seen,
the spherulites of the minor component segregate into cylindrical
shaped domains with an average diameter of 2÷3 μm oriented along
the extrusion direction and uniformely distributed throughout the
whole specimen.

A very drastic change in the overall morphology of these filaments is produced by the addition of a certain amount (10%) of both
copolymers. As can be seen in Fig. 16, the domains of the dispersed

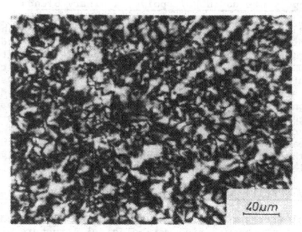

Fig. 8. Optical micrograph (crossed polars) of extrudate cross-
section of pure iPP.

Fig. 9. Optical micrograph (crossed polars) of extrudate cross-
section of pure HDPE.

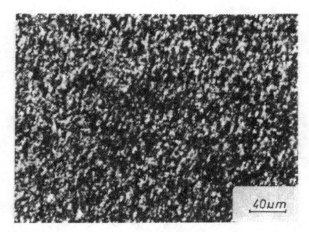

Fig. 10. Optical micrograph (crossed polars) of extrudate cross-
section of iPP/Epcar (70/30) blend.

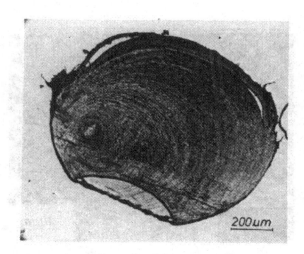

Fig. 11. Optical micrograph (crossed polars) of extrudate cross-
section of HDPE/iPP (25/75) binary blend.

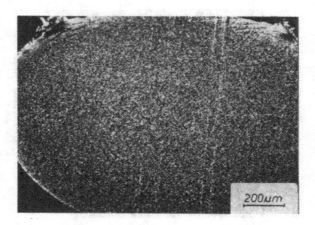

Fig. 12.   Optical micrograph (crossed polars) of extrudate cross-
section of HDPE/iPP/Epcar (23.75/71.25/5) ternary blend.

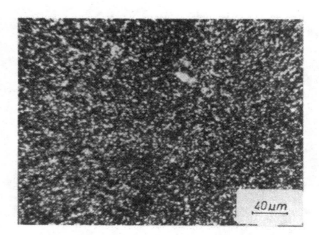

Fig. 13.   Optical micrograph (crossed polars) of extrudate cross-
section of HDPE/iPP (75/25) binary blend.

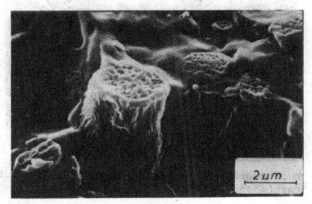

Fig. 14.   SEM micrograph of cryogenic fracture surface of HDPE/iPP
           (25/75) extruded binary blend.

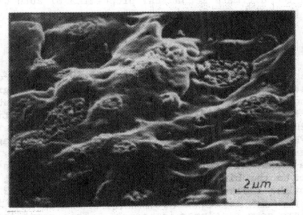

Fig. 15.   SEM micrograph of cryogenic fracture surface of HDPE/iPP
           (75/25) extruded binary blend.

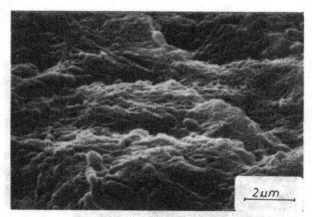

Fig. 16. SEM micrograph of cryogenic fracture surface of HDPE/iPP/
Epcar (65.5/22.5/10) extruded ternary blend.

phase are no longer clearly visible. From the above observations
it may be suggested that EPM copolymers can act as "compatibilizers"
and/or "interfacial agents" for HDPE and iPP. It is likely that
a certain amount of EPM molecules are dispersed in the amorphous
phase of HDPE and iPP giving rise to a more interconnected system.
All these observations are in keeping with the large improvement
observed in the ultimate mechanical properties of such blends as
well as in their impact behaviour[6].

CONCLUSIONS

     The results presented above show that the addition of rubber
ethylene-propylene copolymers (Dutral or Epcar) have a great
influence on the morphology of binary HDPE/iPP blends and consequently on their mechanical tensile properties. In particular,
since the chemical structure of these copolymers is intermediate
between HDPE and iPP, they can act as "compatibilizing agents"
between the amorphous regions of the two semicrystalline homopolymers. As a matter of fact, a better homogenization is achieved
with respect to the binary HDPE/iPP blends. This effect makes it
possible for the internal superstructure of the HDPE/iPP alloys
to release the high stress concentration that can arise in the
material where some copolymer is present. This means that the
copolymer might substantially improve the flow capability of HDPE/iPP
blends so that the cold-drawing process becomes easier and more
continuous as in the case of pure HDPE and iPP homopolymers.

ACKNOWLEDGMENT

This research was supported by a grant from the Progetto
Finalizzato Chimica Fine e Secondaria of the National Research
Council (C.N.R.).

REFERENCES

1. R. E. Robertson, D.R. Paul, J.Appl.Polym.Sci., 17:2579 (1973).
2. R. Greco, G. Mucciariello, G. Ragosta, E. Martuscelli,
    J.Mat.Sci., 15:845 (1980).
3. R. D. Deanin, M.F. Sansone, Polym.Symposia, 19:211 (1978).
4. R. Greco, G. Mucciariello, G. Ragosta, E. Martuscelli,
    J.Mat.Sci., 16:1001 (1981).
5. Ma Rong Tang, R. Greco, G. Ragosta, S. Cimmino, Ibid.
    to be published.
6. L. D'Orazio, R. Greco, C. Mancarella, E. Martuscelli,
    G. Ragosta, C. Silvestre, Polym.Eng.Sci., 22:536 (1982).
7. E. Nolley, J.W. Barlow, D.R. Paul, Polym.Eng.Sci., 20:364
    (1980).
8. L. D'Orazio, R. Greco, E. Martuscelli, G. Ragosta,
    Polym.Eng.Sci., to be published.
9. J. Karger-Kocsis, A. Kallo, A. Szafner, G. Bodo, Z. Senyel,
    Polymer, 20:37 (1979).
10. E. Martuscelli, C. Silvestre, G.C. Abate, Polymer, 23:229
    (1982).
11. B. Maxwell, SPE J., 26:48, June (1980).
12. B. Maxwell, E.J. Dormier, Polym.Eng.Sci., 22:280 (1982).
13. E. Martuscelli, C. Silvestre, R. Greco, G. Ragosta in "Polymer
    Blends Processing, Morphology and Properties", E. Martuscelli,
    R. Palumbo, M. Kryszewski, eds., Plenum Press, New York (1980)
    pp. 295-318.

FORMATION OF EDA COMPLEXES IN POLYMER MIXTURES

Ludomir Slusarski, and Marian Zaborski

Institute of Polymers
Technical University of Łódź
Poland

Heterogeneous mixtures of elastomers and plastomers
are of practical importance and have been widely investigated. As a result of a small difference in solubility
parameter values a diffusion boundary between components
of a mixture or an interfacial chemical bonding between
polymers are formed. It is commonly accepted that good
mechanical properties of such a systems in the two mentioned cases are observed. It is also evident that mixing
conditions and a morphology of a mixture play an important
role. In this work it was rather unexpectedly stated that
the presence of polyacrylonitrile in blends causes strong
increase of viscosity and tensile strength of unsaturated
elastomers. Those effects are clearly related to double
bonds content in elastomer. A lowered segmental mobility
of a matrix in the vicinity of plastomer domains was
observed by means of mechanical spectrometry and pulsed
NMR methods. It is concluded that the charge transfer
mechanism between cyano groups of polyacrylonitrile and
double bonds is probably responsible for that phenomenon.
This hypothesis was confirmed on the basis of IR and UV
spectroscopy examinations of model substances: dodecene-1
and butyronitrile. The constitution of the complex was
established and the enthalpy of its formation calculated.

INTRODUCTION

Heterogeneous mixtures of elastomers and plastomers are widely
used for practical purposes. As a rule, in such systems the elastomer

is a dispersion medium and the plastomer plays the role of an organic filler. A strengthening effect of many plastomers is comparable to that of conventional, mineral fillers, although adequate dimensions of their particles are of one or two orders larger. Two reasons are usually given for the high activity of the organic fillers: a segmental, interfacial diffusion, or grafting[1,2]. It follows from our observations that the influence of polyacrylonitrile on the properties of unsaturated elastomers could not be reasonably interpreted in terms of these assumptions. The goal of our work was to examine the unusual behaviour of such systems with regard to mixtures of other elastomers and plastomers.

EXPERIMENTAL

The objects of our investigations were four kinds of elastomers, of different structure and polarity, viz. cis-1,4-polybutadiene (BR), butadiene-acrylonitrile copolymer (NBR), isobutylene-isoprene copolymers (IIR) and ethylene-propylene-diene terpolymer (EPT). They were mixed with plastomers: low density polyethylene (PE_{1d}), polystyrene (PS), polytetrafluoroethylene (PTFE), polyvinyl chloride (PVC), polycaproamide (PCA) and polyacrylonitrile (PAN) (Table 1). The concentration of the plastomers in the mixtures was changed in the range from 0 to 50 pph of the elastomer. The polymers were blended at temperature T = 423 K by means of the micromill of the Plasti--Corder apparatus. After 24 hours, crosslinking substances, dicumyl peroxide (DCP) or sulphur and diphenylguanidine (S, DPG), were added at room temperature. The composition of the mixtures is given in Table 2.

The viscosity of the uncrosslinked mixture was determined by means of the Mooney apparatus, according to ASTM D 1646-72. The mixtures were vulcanized in cure press with electrical heating at temperature T = 423 K, under pressure P = 30 MPa, during $\tau_{0.9}$, that conforms to the increase of about 90% of maximal shear modulus as measured by Monsanto rheometer. The morphology of vulcanizates was investigated by means of a scanning electron microscope Jeol IA 50 A or a polarisation microscope. To obtain data concerning the adhesion between elastomer and plastomer the surface formed as a result of specimen tearing was examined. The mechanical properties of vulcanizates were measured according to ASTM D 412, using dumb-bell specimens of type B and dynamometer FU 1000e, made in GDR. Thermomechanical analysis of the vulcanizates was carried out by means of semi--automatic apparatus of our own construction[3] in the temperature range T = 153-473 K, the heating rate being 0.034 K/s (2 K/min.). The viscoelastic properties of the vulcanizates in the function of temperature were examined by means of the method of resonance-free, forced vibrations, using Rheovibron apparatus DDV-II-C at frequency of 110 Hz. The mobility of macromolecules and their fragments was examined by means of the nuclear magnetic resonance (NMR) method.

Table 1. The physical parameters of polymers

Polymer	δ	γ_s	X
Elastomers:			
Ethylene-propylene-diene terpolymer, EPT (55% ethylene, 3% DCP)	16.5	32	0.02
Cis-1,4-polybutadiene, PB	17.5	32.5	0.03
Butadiene-acrylonitrile copolymer (33% AN), NBR	20.6	41.5	0.27
Plastomers:			
Polytetrafluoroethylene, PTFE	11.7	26	0.09
Polyethylene, low density, PE_{ld}	16	31.5	0.02
Polyisobutylene, PIB or butyl rubber IIR	17.0	31.5	0.10
Polystyrene, PS	19.1	43	0.10
Polyvinyl chloride, PCV	19.7	42	0.15
Polycaproamide, PCA	25.2	47	0.25
Polyacrylonitrile, PAN	25.7	61	0.76

Solubility parameters δ $\left[MJ/m^3\right]^{0.5}$

Surface tension γ_s $\left[mJ/m^2\right]$

Polymer polarity X = γ_p/γ, γ_p – polar component of the surface tension

Data according to D.W. Van Krevelen, "Properties of polymers", Elsevier, Amsterdam 1976

Table 2. Mixtures composition

Component	Content (phr)					
Cis-1,4-polybutadiene	100	100	–	–	–	–
Butadiene-acrylonitrile copolymer	–	–	100	100	–	–
Ethylene-propylene-diene terpolymer	–	–	–	–	100	100
Zinc stearate	1.5	–	1.5	–	100	–
Diphenylguanidine	1.8	–	1.8	–	1.5	–
Sulfur	2.0	–	2.0	–	0.5	–
Zinc ethyl-phenyl-dithiocarbamate	–	–	–	–	2.5	–
Dicumyl peroxide	–	0.27	–	1.08	2.5	1.35

Plastomer was added to each of compositions in amounts: 0, 10, 20, 30, 40 and 50 (phr)

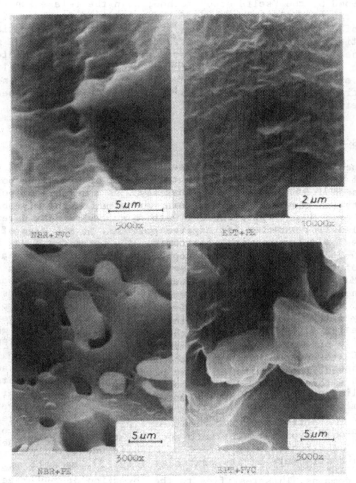

Fig. 1.   Microphotographs of mixtures of elastomers with plastomers
made by means of the electron scanning microscope.

To do this, measurements of relaxation time T_1 (spin-spin) and T_2 (spin-spin) were accomplished by means of a Bruker type SXP 4/100 pulsed spectrometer. Measurements of T_1 were carried out by the method of impulse application Π-τ-$\Pi/2$. Impulse length $\Pi/2$ was 2-3 µs, and field frequency - 90 MHz. The relaxation time T_2 "spin-spin" was determined by the "solid echo" method, when the relaxation time was in the range of 10-200 µs, and if the relaxation time was of the order of 1 ms, the Gill-Meiboom's method[4] was used. Separation of relaxation time components was effected using the graphical method described by Mc Brierty[5]. In order to study the specific interactions between PAN and unsaturated elastomers the mixtures of model substances, i.e. n-dodecene-1, n-dodecane and n-butyronitrile in infrared and ultraviolet were investigated by means of Pye-Unicam spectrophotometer of the SP-700 type.

RESULTS AND DISCUSSION

The dimensions of the dispersed phase particles ranged from several to several dozen µm. From that point of view, the polymeric fillers should be classified as inactive. As a measure of polymer affinity one can accept to a certain aproximation the miscibility parameter $\beta = (\delta_1 - \delta_2)^2$, where δ with a suitable index determines the solubility parameter for a given polymer. In the case of tested mixtures of elastomers with plastomers, the parameter β was in the of 0.25-0.84 MJ/m^3 range. Microscopic examination showed that symptoms of interphase diffusion layer formation appeared when $\beta < 3$ MJ/m^3. In such a case the boundary between polymers was broadened and plastomer domains were firmly embedded in elastomer medium. On the surface of torn vulcanizate specimens the empty spaces usually remaining after the spalling of plastomer domains were not visible (Fig. 1). In the case of the second polymer group, whose miscibility parameter is larger than $\beta > 3$ MJ/m^3, e.g. NBR/PE or EPT/PVC mixtures, no symptoms of the formation of diffusion layer were observed. The domains of these plastomers do not show adhesion towards elastomers (Fig. 1). As could be expected, plastomers, whose solubility parameters are approximately the same as those of the elastomer matrix, behave as active fillers, as, e.g. PE towards EPT (Fig. 2). The activity of two plastomers PTFE and PAN is greater than one could expect taking into consideration the values of parameter β. The strengthening effect of PTFE is undoubtedly due to the anisotropic structure of its particles (Fig. 3), as well as to the mechano-chemical grafting of elastomer macromolecules on the plastomer domains. This testifies to the formation of rubber gel during mixture preparation (Fig. 4).

In spite of the fact that the miscibility parameter of polyacrylonitrile with tested elastomers is included within the bounds of 26-84 MJ/m^3, we did not expect that the polymer would exhibit considerable adhesion towards unsaturated elastomers. However,

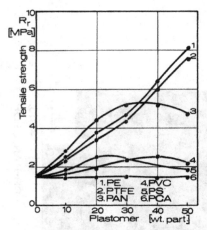

Fig. 2. Influence of plastomers on the tensile strength of EPT vulcanizates.

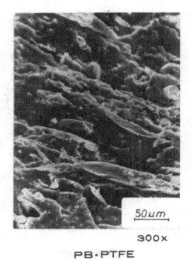

Fig. 3. Microphotograph of PTFE+PB mixture made by means of electron scanning microscope.

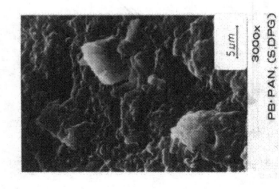

Fig. 5. Microphotograph of PAN+PB mixture made by means of electron microscope.

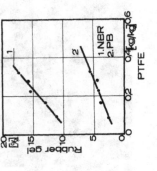

Fig. 4. Rubber gel content in the PB, NBR mixtures with PTFE.

the domains of that plastomer are firmly embedded in the elastomer
medium (Fig. 5). The hypothesis concerning the mechanical grafting
of diene elastomers on polyacrylonitrile domains was rejected, since
no formation of rubber gel was observed in the elastomer mixtures,
which contained PAN. Furthermore, this plastomer is poorly wettable
by the elastomers used; the interphase surface tension calculated
by means of the method proposed by Wu^6 amounts to $\gamma_{PAN,EPT} = 51$,
$\gamma_{PAN,PB} = 50$, and $\gamma_{PAN,KBN} = 27$ mJ/m^2 respectively. It was found
that polyacrylonitrile increased the mixture viscosity all the more
the higher was the concentration of double bonds in the elastomer.
No such relationship was observed in the case of polyethylene
mixtures with butyl rubber (Fig. 6). We came to the conclusion
that the activity of polyacrylonitrile as a filler can be
connected with EDA interactions between electrons of double bonds
and $-C \equiv N$ groups. No such complex type has so far been detected
in polymer mixtures. In the given instance an EDA complex could
appear only at the interphase boundary and its concentration would
be quite low. However, certain symptoms of its existence have been
observed. PAN added in the amount of 30 phr raised the cis-1,4-poly-
butadiene T_g towards higher temperature region by 4-15 K. This was
observed by means of thermomechanical analysis under dynamic as well
as under static conditions (Figs. 7 and 8). The presence of the
immobilized layer of PB on the PAN domains was also established in
studies carried out by the method of pulse NMR. In the mixtures of
PB with PAN there appeared additional compliances of the relaxation
time T_2 "spin-spin" (Fig. 9), as well as relaxation time T_1 "spin-
network" (Fig. 10). This indicates that part of the elastomer has

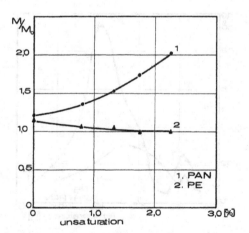

Fig. 6. Dependence of viscosity ratio, M/M_0 on unsaturation of IIR
containing PAN or PE_{1d}. M - viscosity of the mixture,
M_0 - viscosity of the elastomer.

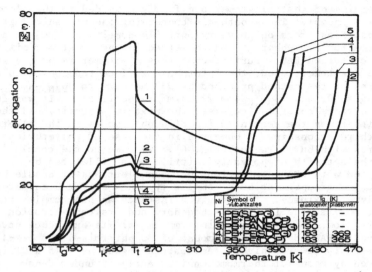

Fig. 7.   Thermomechanical curves of PB vulcanizates containing 30 phr PAN or PE.

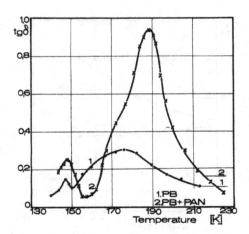

Fig. 8.   The dependence of the mechanical loss tangent of PB vulca-nizates and its mixture with PAN (30 phr) on temperature.

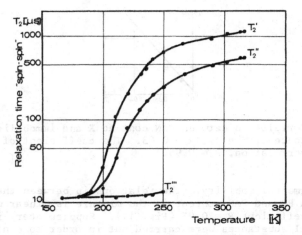

Fig. 9. Relaxation times "spin-spin" of the BR-PAN mixture versus temperature. T_2' - BR, T_2'' - BR immobilized, T_2''' - PAN.

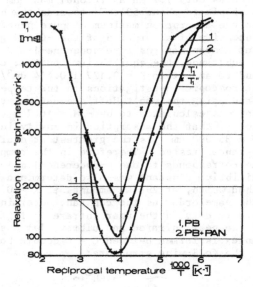

Fig. 10. Relaxation time "spin-network" of the BR vulcanizates containing PAN versus reciprocal temperature.

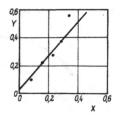

Fig. 11.   Correlation between PAN content X and immobilized BR phase
            content, $Y' = 1.31x - 0.03$. The coefficient of the linear
            correlation, $R = 0.95$.

lowered segmental mobility. The relationship between the content of
immobilized PB and PAN content in the mixture is linear with the correlation coefficient $R = 0.95$ (Fig. 11). Supplementary investigations
on the model substances were carried out in order to confirm the above
hypothesis using n-butyronitrile, n-dodecene-1 and n-dodecane. The
absorption spectra of those substances in the near infrared are
slightly different than the spectra of their mixtures. For instance,
the combination band for the $-C \equiv N$ group in dodecene-1 solution
(dashed curve) (Fig. 12) is slightly different than the curve for pure
substance (continuous curve). An additional maximum of absorption
appears that is displaced about 53 cm^{-1} towards the higher wave num-
bers relative to the absorption maximum of pure n-butyronitrile.
Comparing the intensity of absorption of the $-C \equiv N$ group combination
band in different concentrations into dodecene-1 and dodecane solutions
as well as pure substance, the equilibrium constant of the complex at
298.5 K was found to amount to $K = 0.127 \pm 0.444$ dm^3/mol. On the
basis of IR spectroscopic investigations in the temperature range
298.5 - 333 K, and using the van't Hoff's equation, the enthalpy of
complex formation was calculated to be $-\Delta H = 16.7 \pm 4.3$ kJ/mol. It
was found, moreover, that the absorption intensity in ultraviolet at
wavenumber of $\overline{\nu} = 35\ 000$ cm^{-1} is the greatest at equimolar concentration of the substances tested. Therefore in the complex one n-butyronitrile molecule corresponds to one n-dodecene-1 molecule (Fig. 13).
The complex equilibrium constant was also determined by the Benesi-
-Hildebrand method which, at 298 K, amounts to $K = 0.109 \pm 0.042$ $dm^3/$
mol and is of the same order as the constant determined on the basis
of the spectroscopic data in the near infrared. It follows from
the analysis of the near infrared and ultraviolet absorption spectra
that an EDA complex is formed between $C = C$ double bond and $-C \equiv N$
group which has the following structure:

$$\begin{array}{c} \sim\ C - CH^{\oplus} \sim \\ \int\ \ \ | \\ \sim C - C = N^{\ominus} \end{array}$$

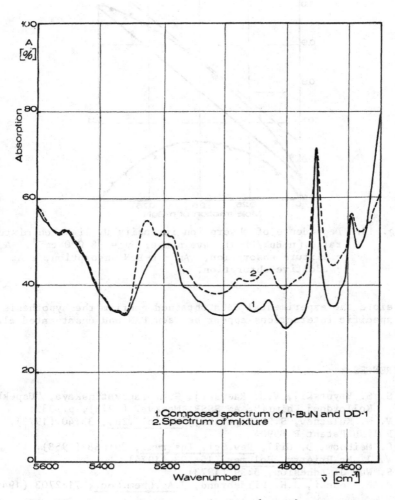

Fig. 12.   IR spectra of n-butyronitrile (n-BuN) and dodecene-1
(DD-1) and their mixture (1:1 by vol.). Thickness of
the layers 2x2 mm.

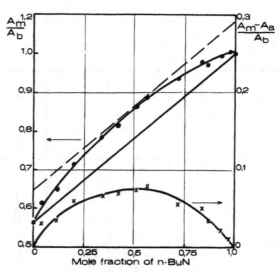

Fig. 13.   Dependence of absorption intensity UV light on mixture
           ratio (n-BuN/DD-1) wave number, $\nu = 35\ 000\ cm^{-1}$. A_m -
           - mixture absorption, A_b - n-BuN absorption, A_a -
           - additive absorption.

Therefore the experimental data obtained confirm the hypothesis that
EDA specific interactions appear between PAN and unsaturated elato-
mers.

REFERENCES

1. S. S. Voyutskij, V.G. Raevskij, S.M. Yagnyatinskaya, "Uspekhi
        Kolloidnoj Khimii", AN SSSR, Moskva, (1973), p. 339.
2. V. N. Kuleznev, S.S. Voyutskij, Kolloid.Zh., 35:40 (1973).
3. Polish Patent P 85904.
4. S. Meiboom, D. Gill, Rev.Sci. Instrum., 29:688 (1958).
5. V. J. Mc Brierty, Polymer, 15:503 (1974).
6. S. Wu, J. Adhesion, 5:39 (1973).
7. H. A. Benesi, J.H. Hildebrand, J.Am.Chem.Soc., 71:2703 (1949).

ELECTRICAL PROPERTIES OF POLYPROPYLYNE-POLYCARBONATE BLENDS

P. Myśliński, Z. Dobkowski and B. Krajewski

Institute of Industrial Chemistry
Rydygiera 8, 01-793 Warsaw, Poland

The electrical properties, i.e. volume resistivity,
dielectric permittivity and dielectric loss factor, as well
as thermally stimulated depolarization current, were measured
on polypropylene-polycarbonate (PP-PC) blends. The results
confirm the existence of some interactions between the non-
-compatible components of PP-PC blends.

INTRODUCTION

Preliminary measurements of mechanical properties of polypropylene (PP) - polycarbonate (PC) blends revealed that the blends
containing about 10 wt.% PP maintain several properties at the level
of pure PC. Some of the properties (e.g. tensile yield, Young's
modulus or impact strength) of these blends are even higher than
those of pure PC [1-3]. We have recently measured the electrical
properties of PP-PC blends, and the results are reported in this
work.

EXPERIMENTAL

Blends of commercial J-330 PP and bisphenol-A PC were prepared
by mechanical mixing of melted components in an extruder at head
temperature of 250° C. The blends were then transformed into 0.5 mm
thick films by compression moulding at temperature 205° C under
pressure up to 20 MPa and samples were cut for measurements of
electric properties.

The following electrical properties of PP-PC blends were

measured:
- volume resistivity, ρ_v, at temperature 20°C, using a Vibron
  Model 62A electrometer; ρ_v was calculated from Ohm's law;
- dielectric permittivity, ε, and dielectric loss factor, tan δ, at
  frequencies 60 and 10^3 Hz and in the temperature range from 20
  to 150°C, using a Tettex test bridge.

Thermally stimulated depolarization (TSD) current was also measured
using an Ekco type 616B electrometer. The samples were polarized at
the following conditions: polarization temperature 150°C, polarization time 2 min, electric field 8 kV/cm. The heating rate was
1°C/min.

RESULTS AND DISCUSSION

        The results are shown in Figs. 1-8. Volume resistivity, ρ_v,
can be considered as the additive property, varying proportionally
to the blend composition. Thus, ρ_v decreases if the content of PC
is increased (ρ_v of PC is lower than that of PP), see Fig. 1.
Dielectric permittivity, ε, is higher for blends of higher PC
content (Fig. 2).

        Dielectric loss factor, tan δ, vs. temperature in the range of
$20-150^\circ$C is represented by similar curves for each blend, (see
Figs. 3 and 4). Tan δ is sharply increased above 130°C, as for pure
PC, and should reach the maximum at the glass transition temperature, Tg, of PC, i.e. at about 150°C. The maximum at 150°C was,
however, not found. Thus, the continuous increase of tan δ should
be attributed not only to the relaxation of polar groups in PC, but
also to the relaxation of free charges (i.e. ions and electrons) in
the blends. Moreover, it was found that tan δ values were the highest
for PP-PC blends containing about 90 wt.% PC in the whole tempera-

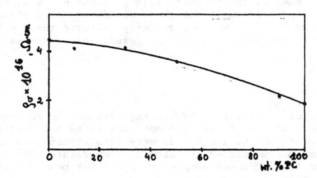

Figure 1.   Volume resistivity, ρ_v, for different PP-PC blend
            compositions.

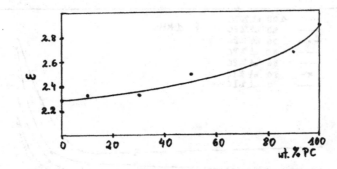

Figure 2. Dielectric permittivity, ε, for different PP-PC blend
           compositions at 20°C and frequency 0.06 KHz.

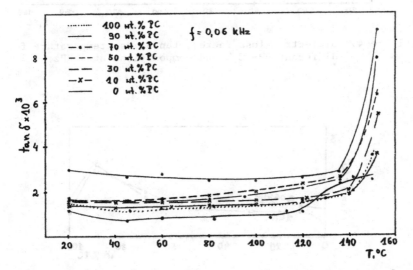

Figure 3. Dielectric loss factor, tan δ, vs. temperature for
           different PP-PC blend compositions at 0.06 KHz.

ture range for both frequencies (see Fig. 5 as an example). The
value of tan δ is related to this part of electric field energy
which is dissipated in a polymer during each cycle of field. This
dissipation of energy depends on the relaxation of polymer polar
groups and on the relaxation of free charges in an alternating
electric field. The higher the content of the polymer polar groups
and its average effective dipole moment, the larger is the dissipation effect. Therefore, the increase of tan δ for PP-PC blends

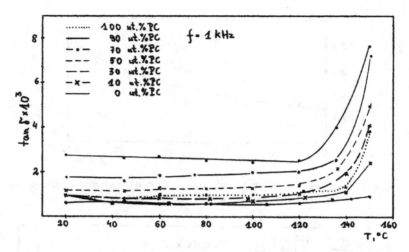

Figure 4. Dielectric loss factor, tan δ, vs. temperature for
           different PP-PC blend compositions at 1 KHz.

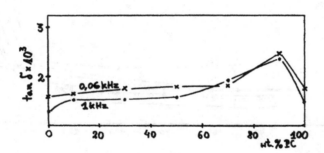

Figure 5. Dielectric loss factor, tan δ, as a function of
           PP-PC blend composition at 20°C.

with 90 wt.% PC can be explained by the increase of average effective dipole moment of PC polar groups.

    The results of measurements of TSD current are shown in Figs. 6
to 8. Maxima at about 65 and 120°C were observed in the measured
temperature range for PP component and at about 150°C for PC component (Fig. 6). The maxima for PP can be connected with the release
of free charge carriers from their traps, depending on the polymer

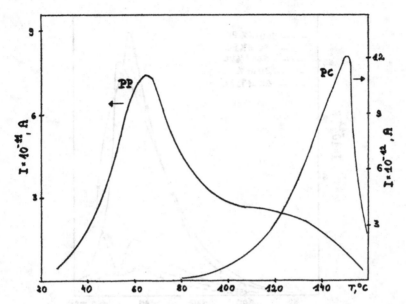

Figure 6. TSD current vs. temperature for PP and PC homopolymers.

morphology [4]. The maximum for PC is connected with the relaxation of polymer polar groups at Tg.

TSD current for PP-PC blends depends, above all, on the PC content, and two maxima of current intensity are observed: the higher intensity maximum at about 150°C attributed to Tg of PC, and the lower one above 150°C. This second maximum can reflect an effect of free charge relaxation which could be connected with the presence of PP in the blends and attributed to the melting temperature of PP. The maximum at 65°C, typical of pure PP, is not observed for PP-PC blends.

The intensity of depolarization current at Tg of a polymer depends also on the content of polymer polar groups and on the ave-rage dipole moment of polar groups [5]. It can be seen from Fig. 7 that the content of PC influences the discharge current intensity. The highest discharge current intensity is observed for the PP-PC blend with 90 wt.% PC. The intensity for 90 wt.% PC blend is even higher than that for pure PC. Thus, the increase of discharge current intensity depends, in this case, not only on the higher content of polar groups, but also on the increase of average dipole moment of these groups.

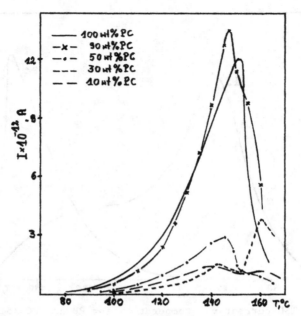

Figure 7.   TSD current vs. temperature for different PP-PC
blend compositions.

Maxima of TSD current, attributed to Tg of PC, Tg_e, vs. the
composition of PP-PC blends, are shown in Fig. 8. The dynamic glass
transition temperature, Tg_d, from measurements of mechanical properties [2], is also ahown in Fig. 8 for the sake of comparison. It is

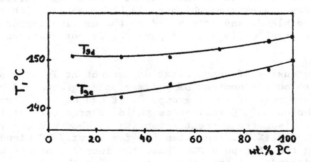

Figure 8.   Maxima of TSD current, Tg_e, and dynamic glass transition
temperature Tg_d [2] vs. PP-PC blend composition.

evident that Tg of PC is decreased in the presence of PP. Therefore, in spite of the heterogeneity of PP-Pc blends, there are some mutual interactions between the components.

Measurements of tan δ and TSD current indicate that there are some characteristic interactions for the PP-PC blend with 90 wt.% PC and the increase of average dipole moment of polar groups is the highest in this case. Some chain orientation of PC macromolecules, induced by the presence of semi-crystalline PP component, can be one of the reasons of such an increase. This assumption, however, was not fully confirmed by x-ray measurements of crystallinity and by electron scanning microscopy of PP-PC blends [3].

CONCLUSIONS

In spite of the visible heterogeneity of PP-PC blends, measurements of the electric properties confirm the existence of some interactions between the components. This finding agrees with the previous measurements of mechanical properties.

REFERENCES

1. Z. Dobkowski, Polimery, 25:110 (1980).
2. Z. Dobkowski, Z. Kohman and B. Krajewski, in "Polymer Blends. Processing, Morphology and Properties" eds. E. Martuscelli, R. Palumbo and M. Kryszewski, Plenum Press, New York, London (1980), pp. 363-372.
3. A. Dems, Z. Dobkowski, B. Krajewski, J. Mejsner, W. Przygocki and H. Szocik, V Conference on Modification of Polymers, Lubiatów, Poland (1980), prepared for publication.
4. P. Myśliński and M. Kryszewski, Polym.Bull., 2:761 (1980).
5. C. Bucci, R. Fieschi and G. Guidi, Phys.Rev., 148:816 (1966).

NEW CONDUCTIVE POLYMERIC COMPOSITES

J. Ulański, J.K. Jeszka and M. Kryszewski

Institute of Polymers
Technical University of Łódź
90-924 Łódź, Poland

A new method has recently been proposed of obtaining
conductive polymeric materials. It consists in crystallizing
a conductive charge-transfer complex during the process of
polymer film casting. If the composition and casting conditions are chosen properly, a conductive material can be
obtained at an additive content as low as 1 wt.%. The paper
considers the influence of different parameters on the formation of a continuous conductive network of CT complex
crystallites. It is shown that the casting temperature, the
solvent used and the type of polymer matrix have considerable
effect on the morphology of CT complex crystallites and consequently on the electrical conductivity of the samples.
For each CT complex - polymer - solvent system there is an
optimum temperature at which the quality of the conductive
network is the best and the conductivity reaches its maximum
value. Above and below this temperature the morphology
changes and conductivity falls down abruptly.

INTRODUCTION

Electrically conductive polymeric composites find many applications. They are usually obtained by mechanically mixing a polymer
with particles of a conductive filler: metal or carbon black. In
such a case a high amount of filler is needed[1]. Conductive charge
transfer complexes were also used to prepare conductive polymer
composites[2], but in the early experiments also a high CT complex
content - sometimes as high as 80% - was necessary to obtain
a conductive material.

In the present work a new method of producing conductive polymeric composites is discussed. In this method the conductive material is obtained as a result of formation of a continuous network of a crystalline, conductive, low molecular weight charge-transfer (CT) complex in polymer matrices[3]. This can be achieved by crystallization of a CT complex during solidification of a polymer matrix in the process of film casting. By choosing appropriate conditions of crystallization one can obtain a huge increase of the conductivity of the mixture containing small amount of CT complex (e.g. conductivity of polycarbonate is increased about 10^{17} times at 1 wt.% tetrathiotetracene-tetracyanoquinodimethane CT complex[4]).

As far as we know, the process of crystallization of low molecular weight substances in polymer matrices has not been studied until now. This problem is complicated because, as is usually the case with multicomponent systems, many factors affect this process. Moreover, small amounts of crystallites in the polymer matrix make most of the direct structural investigation techniques inapplicable. The casting temperature, the properties of the polymer matrix and of the solvent influence the crystallization process and the resultant morphology of the CT complex. On the basis of the results obtained we discuss the conditions which should be fulfilled to obtain low resistivity of the polymer-CT complex systems.

EXPERIMENTAL

The investigated films were obtained by casting onto a glass support, at controlled temperature, a solution containing appropriate amounts of the polymer, the donor and the acceptor in a common solvent: p-dichlorobenzene, chlorobenzene or 1,2-dichloroethane. The weight ratios of the components were such that dry films contained 1 wt.% of the additives, the molar donor-acceptor ratio being 1:1. Film thickness was 15-25 μm. The investigations were carried out on CT-complexes of tetrathiotetracene (TTT) or tetrathiofulvalene (TTF) with tetracyanochinodimethane (TCNQ). Poly(bisphenol-A)carbonate (PC), polyoxyphenylene (PPO), polystyrene (PS), poly(p-chlorostyrene) (PchS), copolymer of polystyrene and chlorostyrene (PSchS), and some mixtures of the above polymers were used as polymer matrices. The electric conduction of the films was measured using a four-probe arrangement in the case of highly conducting samples, or a sandwich-type arrangement in the case of samples with higher resistivity.

RESULTS AND DISCUSSION

Influence of the cast temperature

The cast temperature has significant influence both on the quality of the obtained polymer films and on the crystallization

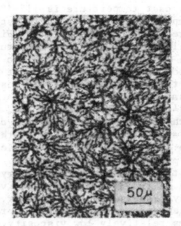

Fig. 1.   Microphotograph of a PC film containing 1 wt.% of TTT·TCNQ.
          Cast temperature 120° C, solvent o-dichlorobenzene.

of additives. For each polymer-solvent system it is necessary to
determine a region of the cast temperatures in which one can obtain
amorphous, transparent, and uniform films. Then, within this region
one should find the optimum temperature, from the point of view of
the formation of crystalline, conductive network. In the case of the
best system known thus far, PC+1 wt.% TTT·TCNQ cast from o-dichloro-

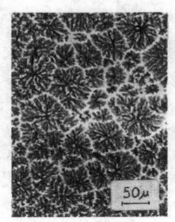

Fig. 2.   Microphotograph of a PC film, containing 1 wt.% of TTT·TCNQ.
          Cast temperature 140° C, solvent o-dichlorobenzene.

benzene, the optimum cast temperature is 115-120° C. The films
obtained at this temperature exhibit the lowest resistivity, about
60 ohm/cm, as measured at room temperature. Fig. 1 shows a microphotograph of the film cast at such a temperature. One can see that
microcrystals of the CT complex are aligned to some extent and form
dendrites growing from nucleation centers. At higher cast temperatures (130-150° C), the neighbour dendrites lose contact and the
resistivity of the samples increases quickly. A microphotograph of
the sample cast at 140° C is presented in Fig. 2. At even higher
cast temperatures, the crystallites disappear due to the lower
complexing constant and better solubility of the additives in the
polymer matrix. The resistivity of such samples is nearly the same
as that of the pure polymer, i.e. 10^{19} ohm/cm. At lower cast temperatures, 80-100° C, the additives form microcrystals with different
morphology. One can see in Fig. 3 that relatively large needle-like
crystals are randomly dispersed in the polymer. This is due to the
fact that at lower temperature the additives precipitate more quickly,
when the solution has relatively low viscosity, and Brownian motions
destroy any arrangement. The slower rate of solvent evaporation permits the growth of bigger microcrystals. It is also possible that at
these temperatures the CT complex has different stoichiometry and,
consequently these crystals have different morphology. The resistivity of the samples increases with decreasing cast temperature e.g.
for films cast at 80° C it is 5×10^3 ohm/cm.

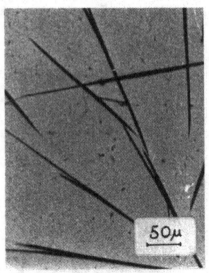

Fig. 3.   Microphotograph of a PC film containing 1 wt.% of TTT·TCNQ.
          Cast temperature 85° C, solvent o-dichlorobenzene.

Effect of the solvent

The solubility of a CT complex in the solvent used has great influence on the crystallization process. If the solubility is too high, the complex does not precipitate at all at any cast temperature at which uniform films can be obtained. This is the case with the system PC+1 wt.% TTF-TCNQ cast from o-dichlorobenzene. When 1,2-dichloroethane was used as the solvent, CT crystallites appeared in films cast at room temperature but their number was still not high enough to form a conductive network. For the complex TTF·TCNQ, due to its relativelly good solubility, we have not yet found any system which would give a conductive material by simple casting. Recent results indicate, that in this case a different procedure of film casting should be used[5].

The solvent can also influence the morphology of the microcrystals formed during film casting, e.g., if chlorobenzene is used as the solvent for the system PC+1 wt.% TTT·TCNQ instead of o-dichlorobenzene, the crystals have the form of long and thin fibres, as can be seen in Fig. 4. The optimum cast temperature in this case is a little lower, c.a. 110° C. Although the morphology of these crystalline structures differs considerably from those obtained from o-dichlorobenzene and described in previous section (Fig. 1), the resistivity of the samples in both cases is nearly the same if cast at optimum temperatures.

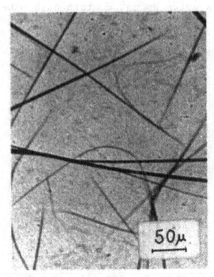

Fig. 4.   Microphotograph of a PC film containing 1 wt.% of TTT·TCNQ.
Cast temperature 105° C, solvent chlorobenzene.

Role of the polymer matrix

Although the polymer matrix does not take part in charge carrier transport which, at least in the first approximation, proceeds along the conductive paths formed by CT complex crystallites, it plays an important role in obtaining the conductive systems. The polymer used must fulfill quite specific conditions for the crystallization process to occur quickly in a relatively high viscosity medium.

The influence of the polymer used as the matrix is exemplified in this section taking TTT·TCNQ as the CT complex, o-dichlorobenzene as the solvent and comparing different polymer matrices.

From the point of view of their utility as matrices for conductive systems containing TTT·TCNQ, the investigated polymers can be ordered as follows: PC, PPO and PS, PS-chS, PchS and PVK.

In the case of PVK, the low molecular weight CT complex does not precipitate at any temperature, because TCNQ forms a competitive CT complex with donor-like carbazole groups. The same was found in the case of PchS and even for copolymer PS-PchS due to the good solubility of the additives in these polymers (monomer units similar to chlorobenzene). For PVK and PchS, even mixing with PC in the ratio 1:2 does not provide good matrices.

In PS films cast at low temperatures the CT-complex forms star--like crystallites which do not contact each other and do not form conductive paths. Near the optimum temperature (90° C in this case),

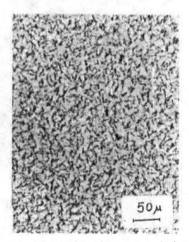

Fig. 5. Microphotograph of a PPO film containing 1 wt.% of TTT·TCNQ. Cast temperature 100° C, solvent o-dichlorobenzene.

TTT·TCNQ forms slightly branched whiskers and the system exhibits a resistivity of 10^3 ohm/cm. In PPO the morphology of crystallized CT complex was also different. One can see in Fig. 5 that very small and thin needles form a fine network, resulting in a resistivity of about 10^3 ohm/cm for the films obtained at optimum temperature (100° C))from o-dichlorobenzene.

Generally speaking, it has been found that the polymer matrix has a significant influence on the crystallization of the complex and the properties of the system. The chemical structure and composition of the polymer used considerably affects the structure of the CT complex crystallites and the optimum cast temperature.

It should be noticed that also the molecular weight of the polymer can influence its properties as a matrix. Experiments on this effect are now in progress.

CONCLUSIONS

The results presented above show that many factors influence the process of crystallization of a CT complex in a polymer matrix. Some of them, like temperature, solvent used, and composition of the polymer matrix, have been discussed in more detail, others, like the molecular weight of the polymer, and evaporation rate have only been mentioned. Investigations are still in progress.

It can be concluded that the high conductivity and other physical properties of these completly organic composites can be controlled by the film casting conditions and composition, depending on the nature of the CT-complex, polymeric matrix, and the solvent used.

It is also note worthy, that under proper conditions crystallization of the CT complex in one direction is strongly favoured yielding films with highly anisotropic conductivity[6].

It seems that these systems give the first possibility of practical application of an interesting and promising new class of materials - one dimensional organic metals.

REFERENCES

1. R. H. Norman, "Conductive rubbers and plastics", Elsevier, Amsterdam (1970).
2. K. Mizoguchi, T. Kamiya, E. Tsuhida and I. Shinohara, J.Polym.Sci. Polym.Chem.Ed., 17:649 (1979).
3. M. Kryszewski, J.K. Jeszka and J. Ulański, Polish Patent P-215189 (1979).

4. J. K. Jeszka, J. Ulański and M. Kryszewski, Nature, 289:90
       (1981).
5. A. Tracz, J. Ulański and M. Kryszewski, Polymer J. (in press).
6. A. Tracz, T. Pakuła, J. Ulański and M. Kryszewski, in this book,
       following paper.

ANISOTROPY OF ELECTRICAL CONDUCTIVITY IN FILMS OF POLYCARBONATE WITH CT COMPLEX OBTAINED BY ZONE SOLIDIFICATION

A. Tracz*, T. Pakuła, J. Ulański* and M. Kryszewski

Centre of Molecular and Macromolecular Studies
Polish Academy of Sciences, Łódź, Poland

*Polymer Institute, Technical University, Łódź, Poland

The aim of this paper is to present a polymeric material showing anisotropy of electrical conductivity obtained by a modified version of a new method of preparing conductive polymeric films.

The method discussed in the paper[1], is based on forming a continuous conductive network of the morphological elements of charge transfer (CT) complex which crystallizes during solidification of the polymeric matrix.

Mechanical deformation, crystallization of the complex in the presence of electrical field, and gradual directional evaporation of the solution of PC with CT complex have been tested as methods to obtain structural modifications of the conductive crystalline network.

Using the latter, so called zone solidification method, anisotropic samples were obtained which are characterized by a conductivity in the conductive direction of the order of 10^{-3} Ω^{-1} cm^{-1} and conductivities ratio of 10^5-10^{10} in the perpendicular directions, at 1 wt.% content of the CT complex in the polymer matrix. The temperature dependences of conductivity in both directions are discussed and related to the morphological structure of the samples.

INTRODUCTION

The mixing of neutral polymers with particles of conductive substances like metals or carbon black[2] for the purpose of modifying the electrical conductivity of polymers has been demonstrated in many works. Electric conduction in such composite materials is achieved due to contacts between the conductive particles dispersed in the polymeric matrix. In order to obtain continuous conductive paths in such a system the quantity of the additive has to be relatively large. However, the presence of these additives has an undesirable effect on the mechanical properties of the composite[3].

A new method of obtaining electrically conductive polymeric materials has been recently proposed by Jeszka et al.[4]. The method is based on forming a continuous conductive network of the morphological elements of charge transfer (CT) complex which crystallizes during solidification of the polymeric matrix. The above authors obtained thin isotropically conducting films of polycarbonate (PC) with only 1% content of CT complex of tetratiotetracene (TTT) and tetracyanoquinodimethane (TCNQ). Specific conductivity of such films is of about $10^{-2} \ \Omega^{-1} cm^{-1}$. Microscopic examination of the films revealed that the basic morphological elements of the CT complex are dendrites which grow up from nuclei randomly appearing in the polymeric matrix during solution thickening. Due to contacts between the crystalline elements belonging to neighbouring dendrites, an isotropic network of crystalline paths is formed. The paths crosslink at dendrite centres. Because of the good electrical conductivity of crystalline CT complex, polymeric matrices containing such a network show relatively high conductivity. The conducting network formed during homogeneous solidification of the sample is isotropic in macroscopic scale and consequently the sample conducts electricity independently on the direction of the applied voltage.

However, the specific structure of the samples suggests that it would be possible to produce samples with anisotropic conductivity if only one were able to modify the morphology in such a way that conductive paths had some orientation or the junctions between conductive elements in isotropic network were broken along some preferred direction.

The aim of the present study was to find methods of obtaining such anisotropically conducting samples. Two approaches were used: modification of the morphology of CT complex in solidified isotropically conducting films and modification of the solidification process in order to obtain the desired effect on the arising morphology.

The first approach which could lead to breaking up of conductive
paths at some preferred direction was effected by mechanical deformation of isotropically conducting films and by treating the conductive
samples with high-voltage electrical pulses. It was expected that
the conductive paths of CT complex can be destroyed in one direction
mechanically or as a result of the Joule-Lentz heat of the electrical pulses.

The zone solidification technique for the oriented growth
of crystals is proposed here as a method of minimizing nucleation
and controlling the growth of conductive crystalline paths of CT
complex during polymer matrix solidification. Basically, the method
consists in passing of a layer of the suitable solution through the
zone with gradient of conditions i.e. temperature, concentration
which controls the processes of solidification and crystallization.
It creates the solidification interface moving throughout the sample
with the velocity dependent on the evaporation rate and usually
leads to remarkable orientation effects in solidified material.

All of the methods above mentioned were tested experimentally
in order to obtain anisotropically conducting samples. However,
satisfactory results have been achieved only by the last method.
This paper will therefore report only the results obtained by the
zone solidification technique.

EXPERIMENTAL

Thin conducting films were cast from solution of 4 wt.% of PC
and TTT TCNQ CT complex in o-dichlorobenzene. The TTT to TCNQ molar
ratio was 1:1, and the CT complex to polymer mass ratio was 0.01.

It was shown in a earlier paper[4] that from such a solution
a good isotropically conducting films can be obtained.

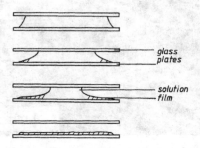

Fig. 1. Schematic representation of film formation by zone
evaporation of the solvent.

     In the present study a version of the zone solidification
method was employed to obtain anisotropically conducting films.
In this method polymer solutions with CT complex were poured between
two parallel glass plates and the solvent evaporated gradually from
the edges of the solution layer. The glass plates were exactly hori-
zontal and the evaporation was performed under normal atmosphere at
120° C. In such conditions a film starts to form on the lower glass
plate at the edges of the solution layer, and the solidification
zone subsequently moves to the inner part. The succesive stages
of the film formation are schematicaly illustrated in Fig. 1.

     The films prepared in this way had the thickness of about 30 μm.

RESULTS

Morphology

     The morphology of the films was examined by means of an optical
microscope. It was established that under the conditions described
above the CT complex starts to crystallize from the nuclei formed
at the edges of the solution layer during the initial stage of
film solidification. As the zone solidification moves throughout
the sample, crystalline columns of the complex, nearly parallel to
each other, are formed at the solidification front. An example
of the structure formed in this way is shown in Fig. 2. It can be
seen from the micrographs that even if a small number of nuclei is
formed at the edge of the sample, a dense system of oriented
crystalline paths is obtained in the inner part of the sample.
The density of branching varies from sample to sample so that films
with various degrees of perfection of crystalline column orientation

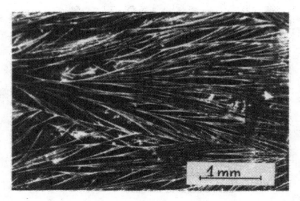

Fig. 2.   Example of the structure of CT complex in film obtained
          by zone solidification.

are obtained. This is probably due to difficulties in preserving exactly reproducible conditions of film preparation. Details of the morphological structure of films are sensitive first of all to the temperature of the solution and of the surroundings, the composition of the solution, and the distance between glass plates.

Electrical conductivity

The electrical conductivity of these films was locally measured by the four electrode method in two directions: along the direction of the conducting path and perpendicular to it. The samples with lower conductivity values were measured using the conventional two electrode arrangement. Values of conductivity determined at room temperature along the growth direction of crystalline columns were very similar for all samples and were of the order of 10^{-3} $\Omega^{-1}cm^{-1}$ while in the perpendicular direction conductivity values were strongly dependent on the density and branching frequency of crystalline paths. The conductivity values obtained were scattered within the range of 10^{-8} $\Omega^{-1}cm^{-1} \div 10^{-13}$ $\Omega^{-1}cm^{-1}$. This means that anisotropy of conductivity measured as the ratio of conductivity values in two perpendicular directions ranges from 10^5 to 10^{10}.

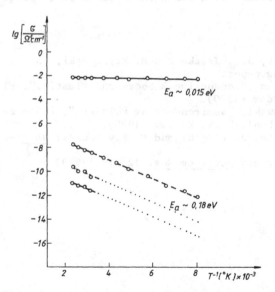

Fig. 3.   Temperature dependence of electrical conductivity:
(——·——)   along the conduction paths,
(— — —)   perpendicular to conduction paths.

Such anisotropy has macroscopic character and should be related
to the morphology of the whole sample and not to the anisotropy of
conductivity of CT complex crystallites themselves. The hypothesis
that conductivity of the investigated materials is determined by
junctions between CT complex crystallites is further confirmed by
measurements of the temperature dependences of conductivity in both
directions.
One can see from Fig. 3 that conductivity in the direction parallel
to the crystal columns is nearly temperature independent being not
far from the temperature dependence of conductivity found for "meta-
lic" CT complexes[5]. The conductivity in perpendicular direction has
considerably higher activation energy because it is limited by the
conductivity of the polymer matrix which separates crystal columns.
These results also show that the anisotropy of conductivity of the
samples obtained is temperature dependent.

CONCLUSIONS

    A method has been developed which makes it possible to produce
conductive polymeric composites with significant anisotropy of conductivity. Due to high ratio of conductivities in two perpendicular
directions the materials obtained can be treated as linear polymeric
conductors. Further work along these lines is in progress.

REFERENCES

1. J. Ulański, J.K. Jeszka and M. Kryszewski, in this book, pre-
        ceeding paper.
2. R.H. Norman, "Conductive Rubbers and Plastics", Elsevier,
        Amsterdam (1970).
3. M. Kryszewski, "Semiconducting Polymers", PWN - Polish Scien-
        tific Publishers, Warsaw (1980).
4. J.K. Jeszka, J. Ulański and M. Kryszewski: Nature, 289:390
        (1981).
5. J.B. Torrance, Acc.Chem.Res. 12 : 79 (1979).

THE INFLUENCE OF INTERFACIAL GRAFTING ON THE PROPERTIES

OF ELASTOMER - PLASTOMER MIXTURES

Ludomir Slusarski and Jerzy Kuczyński

Institute of Polymers
Technical University of Łódź
Poland

INTRODUCTION

Plastomers are often used for modification of elastomers, and they usually form a discrete phase. As could be expected, interfacial grafting should improve the properties of such heterogeneous systems, but there are many difficulties in effecting this reaction in bulk. Under the influence of an initiator, an elastomer as a more reactive component of the mixture is first of all crosslinked. This report presents the results of our attempts to increase the probability of grafting.

EXPERIMENTAL

Our investigations involved four kinds of rubbers: natural (NR). ethylene-propylene (EPR), cis-1,4-polybutadiene (PB) and butadiene--acrylonitrile (NBR). They were mixed with plastomers, polypropylene (PP), polystyrene (PS) and modified polystyrene containing peroxide (PSO), dithiocarbamate (PST) and perthiocarbamate (PSTS) groups.

The following methods of polystyrene modification were employed. The peroxide groups were introduced into macromolecules of atactic polystyrene, with initial viscosity average molecular weight, $\overline{M}_v$ = = 107000 (PS$_{107}$) in the oxidation reaction. The solution of PS$_{107}$ in toluene was bubbled with oxygen at elevated temperature T = 353 K. The reaction of addition of 2,2'-azo-bisisobutyronitrile (AZBN) was initiated by UV irradiation. Polystyrene containing organosulphide groups was obtained by polymerization of styrene in the presence of tetramethylthiuram disulphide, or tetramethylthiuram disulphide and sulphur in the light of a UV lamp. The reaction temperature was

179

T = 443 K. Low molecular weight polystyrene was prepared by thermal
polymerization of the monomer in tetrahydronaphthalene (Tetralin),
under the conditions described in the work of Mayo[1]. The reaction
temperature was 403 K. The molecular weight of oligopolystyrene
could be varied by changing the molar styrene/tetralin ratio. The
polymers were blended at temperature T = 423 K by means of the
micromill of the Plasticorder apparatus. The concentration of plastomers in the mixtures was changed from 0 to 50 pph in relation to
the elastomer. The mixtures were vulcanized in press with electrical
heating at temperature T = 423 K, under pressure P = 30 MPa, during
$\tau_{0.9}$.

The vulcanizate obtained had shear modulus as measured by the
Monsanto rheometer equal to 90% of the maximal value. The mechanical
properties of vulcanizates were measured according to ASTM D 412
using dynamometer FU 1000 e (GDR). Thermomechanical analysis of the
vulcanizates was carried out by means of semi-automatic apparatus
of our own design in the temperature range of T = 153-473 K, the
heating rate being 1 K/min. The morphology of the vulcanizates was
investigated by means of a scanning electron microscope Jeol IA 50A.
DTA analyses were carried out using a Derivatograph apparatus (Hun-
gary). The heating rate was 7.9 K/min, sample weight 60 mg, and the
temperature ranged from 293 to 1073 K. The activation energy of
polymer destruction was calculated by the Freeman-Carrol method[2].
The structure of modified polystyrene was examined by the nuclear
magnetic resonance method (NMR, Tesla apparatus type BS 487 C),
infrared spectroscopy (SP-200 Pye-Unicam IR spectrophotometer), mass
spectrometry (CGMS 2091 LKB spectrometer), and elementary analysis.

RESULTS

At first, the method of "dynamic covulcanization" was applied
to mixtures of natural rubber with polystyrene and ethylene-propylene
rubber with propylene respectively. 100 wt. parts of the components
used in the ratio of 10:1 were mixed with 4 wt. parts of discumyl
peroxide, at shear rate $\dot{\gamma}$ = 91.2 s^{-1}. The temperature of mixing was
T = 423 K (polystyrene) or 453 K (polypropylene). Owing to the use
of small amounts of elastomers in the first step of modification,
the reaction with dicumyl peroxide proceeded practically only in the
diffusion layer. Under these conditions the probability of grafting
was enhanced. Indeed, NMR investigations of the fractionated samples
indicated that elastomers were grafted on the plastomers. Following
that we made mixtures of modified polystyrene with natural rubber and
modified polypropylene with ethylene-propylene rubber. 10 to 50 weight
parts of plastomers were added per 100 weight parts of elastomers.
The appropriate amount of dicumyl peroxide was added and the second
step of "dynamic covulcanization" took place. In this way vulcanizates
with rheological properties, similar to those of SBS triblock copolymers, were obtained (Fig. 1). Unfortunately, their mechanical

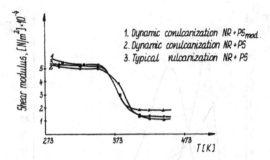

Fig. 1. Shear modulus of natural rubber (NR) mixtures with 50 wt.%
         of polystyrene (PS) as a function of temperature.

properties were not so good as those of SBS copolymers (Fig. 2).
Probably the elastomer grafted in the first step onto the plastomer
was highly crosslinked and, as a consequence, its reactivity toward
the second portion of the same elastomer was lowered.

     To solve the problem of grafting we decided to introduce first
reactive functional groups into the polystyrene macromolecules, e.g.
peroxide, dithiocarbamate, and perthiocarbamate groups. Polystyrene
containing peroxide groups was obtained by oxidizing a commercial
polymer sample according to the method described earlier. The dependence of peroxide groups concentration on oxidation time was of
extreme character (Fig. 3). Initially, the formation of peroxide and
hydroperoxide groups prevailed. At the same time the molecular weight
of polystyrene quickly decreased. When the initiator is consumed,
the reaction of peroxide groups decomposition begins to prevail.

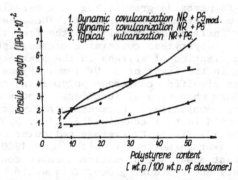

Fig. 2. The influence of polystyrene PS contents on the tensile
         strength of natural rubber (NR) vulcanizates.

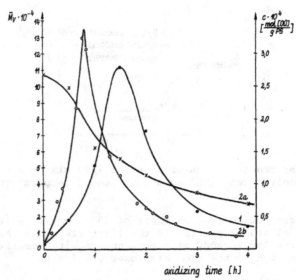

Fig. 3.   The changes of peroxide groups concentration (curve 1)
          and viscosity average molecular weight $\overline{M}_v$ (curves 2a, 2b)
          during the oxidation of polystyrene. 2a - integral curve,
          2b - differential curve.

In the last stage of the reaction a small amount of gel was formed.

     In subsequent investigations oxidized polystyrene was used with
viscosity average molecular weight $\overline{M}_v$ = 40.000 containing
$3 \cdot 10^{-4}$ mol|OO|/g·PS peroxide groups, e.g. one peroxide group per
about 30 styrene monomeric units (Table 1). These peroxide groups
are reactive; for instance, they initiate the free radical polymerization of styrene. Polystyrene containing organosulphide groups was
obtained by polymerization of styrene in the presence of tetramethylthiuram disulfide in the light of an UV lamp. It was found that
tetramethylthiuram disulfide can cause polymerization of styrene
but it was not clear what type of functional groups appeared at
the ends of the macromolecules[3]. Elementary analysis of modified
polystyrene indicated that it contained 1.01% N, 5.01% S, 85.32% C
and 7.65% H (Table 1). Viscosity average molecular weight was $\overline{M}_v$ =
= 1820 and number average molecular weight $\overline{M}_n$ = 1920. From these
data it follows that about 40% of macromolecules contain two thiuram
end groups and only one residual group. This would be the case if
dithiocarbaminian radicals were the only factor initiating the
polymerization of styrene. We supposed that some macromolecules

Table 1. The properties of the modified polystyrene

Symbol	The kind of functional groups	Concentration	Molecular weight
PSO	$-O_{.}O-$	$3 \cdot 10^{-4} \frac{mol\ [OO]}{g\ PS}$	40000
PST	$(CH_3)_2-N-C-S-$ $\overset{\|}{S}$	1.01% N; 5.01% S	1920
PSTS	$(CH_3)_2-N-C-S_x-$ $\overset{\|}{S}$	1.07% N; 7,91% S	2080
PS$_3$	unmodified	–	3120
PS$_{40}$	unmodified	–	40000

were terminated by groups other than dithiocarbaminian. This hypothesis was confirmed by the results of derivatographic investigations.
On the derivatograms of modified polystyrene two peaks are visible
in the low temperature region, probably connected with splitting off
of the end groups. It follows from the thermogravimetric and DTA
curves that at the temperature of 212° C in air, only 58% of the
initiator residue split off, the remaining 42% being firmly attached
to the polymer chain (Fig. 4). It does not split off even at 280° C.
The activation energy of the destruction of thiuram modified polystyrene, E = 23.7 kcal/mol, is a little smaller than the activation
energy of unmodified polystyrene destruction, with approximately the
same molecular weight ($\overline{M}_v$ = 3120), E = 26 kcal/mol. On the basis of
infra-red and nuclear magnetic resonance spectroscopy investigations ·
we discovered the presence of functional groups in modified polymer
but it was difficult to recognize their structure.

To solve this problem we used the mass-spectroscopy method.
The structure of the ion fragments indicated that dithio- and
perthiocarbamate groups were present at the chain ends, but the
high concentration of $(CH_3)_2N-C^{\oplus}$ ions suggested that polystyrene
$\overset{\|}{S}$
macromolecules also contained thiocarbamate groups (Fig. 5). The
structure of polystyrene modified by thiuram was also as follows:

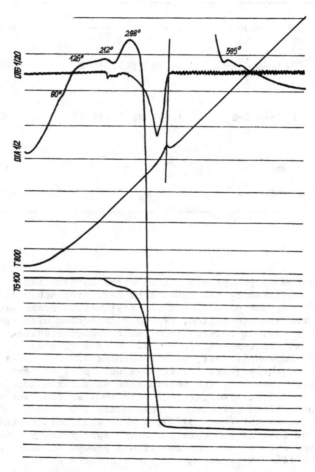

Fig. 4.   Derivatogram of polystyrene modified by thiuram.

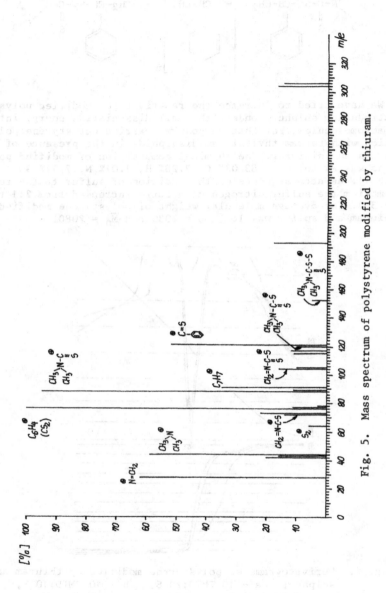

Fig. 5. Mass spectrum of polystyrene modified by thiuram.

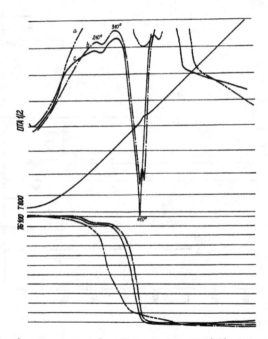

x, y = 1,2

We have tried to increase the reactivity of modified polystyrene
by introducing sulphur bonds with small dissociation energy into
its macromolecules. For that purpose we carried out styrene polymerization with tetramethylthiuram disulphide in the presence of
sulfur as an initiator. The chemical composition of modified polystyrene was as follows: 83.02% C, 7.28% H, 1.07% N, 7.91% S
(Table 1). As acconsequence of the addition of sulfur to the reaction medium the sulfur/nitrogen mole ratio increased from 2:1 to
above 3:1. The average molecular weight of polystyrene modified
by thiuram and sulfur was low, $\overline{M}_v$ = 2030 and $\overline{M}_n$ = 2080.

Fig. 6.   Derivatograms of polystyrene modified by thiuram and
          sulphur;   a - 10 TMTD:20 S,   b - 10 TMTD:10 S,
          c - 10 TMTD: 5 S.

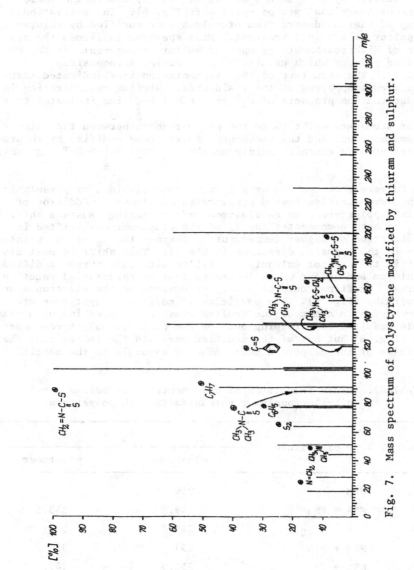

Fig. 7. Mass spectrum of polystyrene modified by thiuram and sulphur.

Derivatographic examination indicated that in the case of polystyrene modified by thiuram and sulfur at 210° C, nearly the whole
amount of functional groups split off (Fig. 6). The activation
energy of thermal destruction of polystyrene modified by thiuram
and sulfur is E = 19.1 kcal/mol. Mass spectrum confirmed the presence of dithiocarbamate groups and sulfur arrangements in the
modified polymer which create $S_2^{\oplus}$ ions during decomposition
(Fig. 7). The structure of the fragmentation ions indicated that
these groups appeared at the chain ends. Similar concentration in
the destruction products of $S_2^{\oplus}$ and $(CH_3)_2N-\overset{\oplus}{\underset{\parallel}{C}}$ ion indicated that
$\qquad\qquad\qquad\qquad\qquad\qquad\qquad\qquad\qquad\qquad\qquad S$
sulfur atoms are built in to the polymer chain between the thiocarbamate groups and the backbone. Polystyrene modified by thiuram
and sulfur also contains mainly unstable $(CH_3)_2N-\underset{\parallel}{C}-S-S-^{\oplus}$ groups.
$\qquad\qquad\qquad\qquad\qquad\qquad\qquad\qquad\qquad\qquad\qquad\qquad\qquad\qquad S$

Polystyrenes containing peroxy, organosulphide, or persulphide
groups were reactive toward unsaturated elastomers. Addition of
modified polystyrene to an elastomer after heating causes a shift
of T_g of both components; the T_g of the elastomer was shifted in
the direction of higher temperatures, whereas the T_g of the plastomer - in the opposite direction (Table 2). This shift is most distinct in the case of vulcanizates filled with polystyrene modified
by thiuram and sulfur which indicates that an interfacial reaction
occured. Grafting could be easily recognized on the electron scaning
micrographs (Fig. 8). The particles of modified polystyrene are
surrounded by a sheet of the grafted elastomer. Breaking of a mixture
sample resulted in "dropping out" of the ordinary polystyrene particles (Fig. 8) but not of the modified one. In the latter case the
particles of the dispersed phase adhered strongly to the matrix.

Table 2. Glass transition of components in butadiene acrylo-
          nitrile copolymer vulcanizates with polystyrene

No.	Type of mixture	T_g K	
		elastomer	plastomer
1.	NBR	249	–
2.	NBR + PS$_3$	249.5	353.5
3.	NBR + PS$_{40}$	248.5	359
4.	NBR + PSO	251	356
5.	NBR + PST	253.5	347.5
6.	NBR + PSTS	256	337

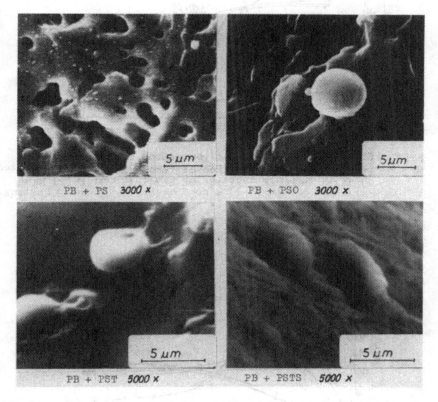

Fig. 8.   Electron scanning micrographs of cis-1,4-polybutadiene
mixtures with modified polystyrene.

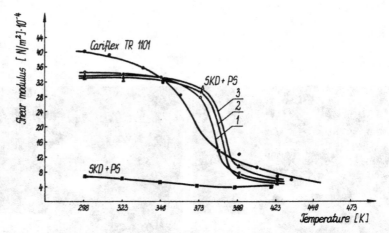

Fig. 9.   Shear modulus of the polybutadiene mixture with oxidized
          polystyrene as a function of temperature;
          1) first heating,   2) second heating,   3) third heating.

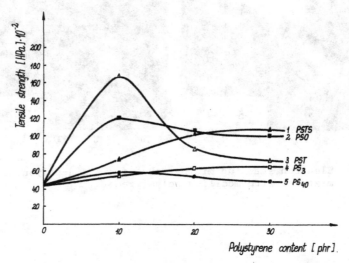

Fig. 10. The influence of polystyrene on the tensile strength
          of the butadiene-acrylonitrile copolymer vulcanizates.

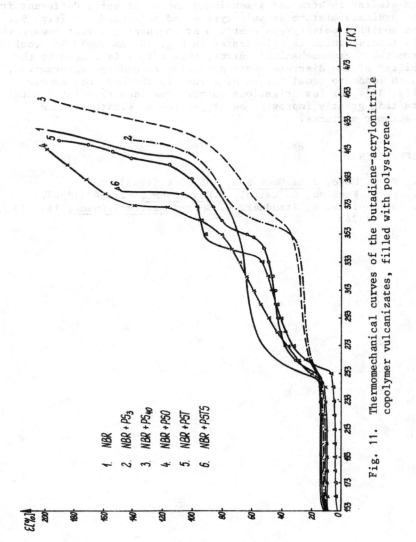

Fig. 11. Thermomechanical curves of the butadiene–acrylonitrile copolymer vulcanizates, filled with polystyrene.

1. NBR
2. NBR + PS₃
3. NBR + PS₄₀
4. NBR + PS₀
5. NBR + PST
6. NBR + PST5

It was found that mixtures of modified polystyrene with elastomers, e.g. cis-1,4-polybutadiene, behave like the SBS triblock copolymers. The changes of the shear modulus as a function of temperature
are similar in both cases mentioned above but quite different from
the ordinary mixture of polystyrene and polybutadiene (Fig. 9).
The modified polystyrene exerts a strengthening effect toward the
elastomers, which is illustrated in Fig. 10. As could be concluded
from the thermomechanical curves, this effect is caused by the
ability of the discrete phase particles to undergo deformation,
which leads to equalization of stress in the deformed sample
(Fig. 11). Our investigations support the idea that interfacial
grafting greatly improves the properties of elastomer and
plastomer mixtures.

REFERENCES

1. F. R. Mayo, J.Am.Chem.Soc., 65:2324 (1943).
2. E. S. Freeman, B. Carrol, J.Phys.Chem., 62:394 (1958).
3. J. Beniska, E. Staudner, E. Spirk, Makromol.Chem., 161:113
       (1974).

SYNTHESIS AND CHARACTERIZATION OF INTERFACIAL AGENTS

TO BE USED IN POLYAMIDE 6/RUBBER BLENDS

M. Avella, N. Lanzetta, G. Maglio, M. Malinconico,
P. Musto, R. Palumbo and M.G. Volpe

Istituto di Ricerche su Tecnologia dei Polimeri
e Reologia, C.N.R., Arco Felice (Napoli), Italy

INTRODUCTION

Polyamides PA have poor impact resistance at temperatures below T_g particularly in the dry state. This inconvenience can be eliminated by incorporating in polyamide matrix a rubbery toughening agent. To improve the adhesion between the phases, minimize the interfacial energy, and influence the dimension of the dispersed rubber particles, a third component, such as a graft or block copolymer of appropriate chemical structure, can be added to such PA/rubber blends. These copolymers are generally referred to as "interfacial agents"[1]. Suitable interfacial agents for PA/rubber blends can be preformed and then added to the homopolymers to be blended or can be formed in situ by chemical reactions during the mixing process. In both cases it is necessary to modify the rubbery polymer in order to introduce appropriate functional groups along the chain[2-6]. In this paper we report on the preparation of a functionalized ethylene-propylene random copolymer (EPM) and on the synthesis of EPM-g-PA 6 graft copolymers utilizing the functionalized EPM. Preliminary results on the use of modified EPM rubbers in PA 6/EPM blends have been previously reported[6].

EXPERIMENTAL

Materials and equipment

Benzoyl peroxide (BPO) was dried on phosphorous pentoxide under vacuum. Commercial ε-caprolactam (CL) was dried on phosphorous pentoxide and twice recrystallized from cyclohexane. Xylene was dried by refluxing over sodium metal. Maleic anhydride (MAH) and EPM

193

(Dutral Co54 by Montedison SpA, 53% of C_2 by weight) were used as received. N-acetylcaprolactam (AcCL) was distilled before use.

The thermal properties were investigated by differential scanning calorimetry on a Perkin-Elmer DSC-2 calorimeter. The melting temperatures and the apparent melting enthalpies were determined from the maxima and the area, respectively, of the melting endotherm by comparison with those of pure reference standards.

Determinations of grafted anhydride groups were carried out on the automatic titration assembly Titroprocessor produced by Metrohm. A glass electrode and a silver chloride electrode were employed as reference. Lithium chloride saturated solutions in ethanol and iso-propanol were used as inner and outer electrolytes respectively.

Preparation of EPM-g-succinic anhydride (EPM-g-SA)

In a three-necked round bottom flask equipped with nitrogen inlet, refrigerator and separating funnel, 5.0 g of EPM were dissolved by refluxing in 100 ml of dry xylene. MAH, 5.76 g, and, through the funnel, a solution of BPO (0.50 g) in 15 ml of dry xylene were added and the solution was refluxed for 15 min. After cooling, the solution was poured in 1 liter of dry acetone, and the resulting precipitate was collected and washed repeatedly with 100 ml portions of acetone and finally dried under vacuum at 50° C. A quantitative evaluation of the amount of grafted anhydride groups was performed by potentiometric titration as follows. Fully hydro-lyzed EPM-g-SA (0.258 g) was dissolved in 50 ml of an o-dichloro-benzene/ethanol solution 9/1 v/v and titred with a 8.5×10^{-2} M solu-tion of potassium hydroxide in o-dichlorobenzene/ethanol 9/1 v/v. The titration gave two inflection points corresponding to the two carboxylic groups, resulting in a grafting value of 2.9% by weight.

Preparation of EPM-g-hexylsuccinimide

In a three-necked round bottom flask with nitrogen inlet, refri-gerator and separating funnel, 1.08 g of EPM-g-SA (2.9% by weight of SA) were dissolved in 50 ml of dry THF. A solution of 1.8 ml of hexylamine in 15 ml of THF was added through the funnel and the mixture was refluxed for 30 min. After cooling, the solution was poured into dry acetone and the precipitate was washed several times with acetone and dried under vacuum. The IR spectrum performed on solution cast films shows amide bands at 1640 and 1545 cm^{-1} and a C = 0 stretching at 1710 cm^{-1}. A portion of this product was heated at 150° C for 20 min and two strong imide bands at 1700 and 1775 cm^{-1} appeared. Elemental analysis gave 0.36% N (calc'd. 0.38).

Typical preparation of a EPM-g-PA 6 graft copolymer

EPM-g-MS (1.0 g, 6.5% w/w of MS, 0.5 mmole) and CL (5.70 g, 50 mmole) were placed under nitrogen atmosphere in a dry three-necked glass cylinder fitted with a mechanical stirrer, nitrogen inlet, and refrigerator. Dry xylene (10 ml) was added and the mixture was refluxed under stirring until a homogeneous solution was obtained. To this solution, kept at 139° C, 1 ml of 0.5 M solution of AcCL in xylene and 1 ml of 3 M solution of NaCL in molten CL were added. After three hs the reaction mixture was cooled to room temperature, 50 ml of methanol were added and stirring was continued for another 15 min. The mixture was filtered and the residue was finely ground, Soxhlet extracted with methanol and finally dried, yielding 5.70 g of a PA 6/copolymer mixture (73% of CL conversion). PA 6 was removed from this mixture by several formic acid extractions leaving 1.85 g of copolymer. Elemental analysis gave 5.8% N, corresponding to 46% by weight of PA 6. The formic acid extracts were evaporated yielding 3.80 g of pure PA 6 as confirmed by IR and H^1 NMR analyses (η_{inh} = = 0.60 dl/g in m-cresol, 25° C, c = 0.5 g/dl).

In the bulk polymerization experiments, xylene was removed by distillation before adding AcCL and NaCL.

RESULTS AND DISCUSSION

Synthesis of EPM-g-succinic anhydride (EPM-g-SA)

The choice of maleic anhydride (MAH) as the functionalizing agent is suitable for several reasons. The most important one - - a fact which distinguishes MAH from other unsaturated molecules bearing functional groups - is that MAH, similarly as other 1,2-di-substituted olefins, does not homopolymerize easily. This makes the grafting product, EPM-g-SA, practically unmodified in its rubbery properties and miscible with the parent EPM.

The grafting reaction of MAH onto EPM was performed in refluxing xylene solutions by means of BPO as radical source[6,7]. The time-dependence of the weight % of grafting is shown in Fig. 1. The reaction is fully accomplished within 15 minutes, which is consistent with the decomposition rate of BPO at the reaction temperature in aromatic hydrocarbons. We assume that primary radicals R· derived from the BPO splitting decomposition are likely to abstract a hydrogen atom from the EPM backbone, leading to P· macroradicals. The addition of a MAH molecule to P· will, in turn, give a new macroradical P· which, by transfer (t) and/or combination (c) reactions, ends the radical sequence without side chain polymerization:

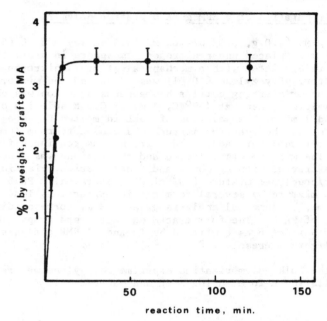

Fig. 1.   Dependence of the weight % of grafted maleic anhydride (MA)
           on the reaction time (EPM 5.0 g; xylene 100 ml; MA 5.75 g;
           BPO 0.50 g).

    In Figure 2 the dependence of the percentage of grafting on the
BPO concentration is plotted. The initiator efficiency is always less
than one, and decreases as the BPO concentration increases, especially at higher BPO concentrations. This finding may result either
from a decrease of reactive sites on EPM and/or from an increased
amount of macroradicals used up in chain scission reactions. Viscosity measurements on EPM before and after the MAH grafting showed
a lowering of molecular weight with increased MAH content, thus
confirming the latter hypothesis.

    Fig. 3 shows the dependence of the percentage of grafting on
the MAH concentration. As the BPO concentration is kept constant,

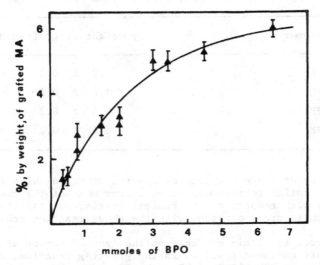

Fig. 2. Amount of grafted maleic anhydride, % by weight, against
BPO used (EPM 5.0 g; xylene 100 ml; MA 5.75 g).

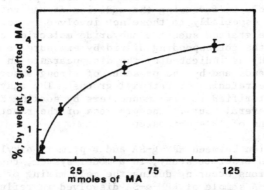

Fig. 3. Variation of grafted maleic anhydride MA , % by weight,
on the MA concentration (EPM 5.0 g; xylene 100 ml; BPO
0.50 g).

Table 1. Comparison of different amounts of maleic anhydride
          grafted onto LDPE, IPP, aPP and EPM

Polymer	% by weight of grafted MAH
LDPE	7.0 ± 0.2
aPP	1.2 ± 0.2
iPP	0.4 ± 0.2
EPM	6.6 ± 0.2

the increase of the percentage of grafted anhydride with increasing
MAH concentration corresponds to an improvement of the reaction
efficiency with respect to the radical initiator. All the results
presented here are consistent with the above reaction scheme.

    In order to obtain evidence on the type of carbon atoms of the
polyolefinic backbone involved in the grafting reaction, we carried
out grafting on EPM, LDPE, i-PP and a-PP under the same experimental
conditions. The results, collected in Table 1, show a very low
degree of grafting for a-PP and i-PP suggesting a preferential attack
of MAH to the secondary carbon atoms of EPM. Secondary and tertiary
macroradicals, and consequently their relative populations, differ
from each other both with regard to the ease with which they form
and their reactivity. In our opinion, steric effects – enhanced
by the bulkiness of the entering MAH molecule – possibly also play
an important role in favouring the addition of MAH to secondary
macroradicals, especially to those not involved in polypropylene
sequences. The grafted succinic anhydride molecules can be easily
hydrolized to the corresponding diacid by exposure to moisture for
one day (50% RH) as indicated by the disappearance in the IR spectrum
of anhydride bands and by the presence of strong absorption at
1710 cm^{-1} (C=O stretching of carboxyl groups). The anhydride molecule
can also be esterified to give monoesters by soaking EPM-g-SA in
alcohols for several hours. The progress of the reaction was
checked by means of IR spectroscopy.

    The reaction between EPM-g-SA and a primary alkylamine was
investigated in order to study, in a model system, the possible
chemical reactions occuring during the melt mixing of Polyamide 6
and EPM-g-SA. A sample of EPM-g-SA dissolved in refluxing THF was
therefore reacted with a large excess of hexylamine. IR analysis
of the reaction product indicated that at relatively low temperatures amidation occurred. By heating a sample of this product at
150° C, progressive disappearance of amide bands and growth of imide

bands was observed. Elemental analysis indicated that practically all the grafted anhydride groups had reacted with the amine. These results suggest that $-NH_2$ end groups of PA 6 react with anhydride groups of functionalized EPM yielding EPM/PA 6 graft copolymers during the melt mixing of EPM, PA 6 and EPM-g-SA at 260° C[6]. At this temperature formation of imide linkages between PA 6 side chains and EPM backbones is likely to occur.

Synthesis of EPM-g-PA 6 graft copolymers

High molecular weight Polyamide 6 can be prepared by means of activated anionic polymerization of ε-caprolactam (CL) at temperatures lower than the melting point of PA 6[8,9]. Graft copolymers with polyolefinic backbones and polyamidic side chains were prepared by this procedure utilizing polyolefins bearing appropriate functional groups capable of promoting the anionic polymerization of CL[3,5]. We therefore synthesized EPM-g-PA 6 graft copolymers by caprolactam polymerization in the presence of sodium caprolactam (NaCL) as the initiator and EPM-g-SA or its methyl emiester (EPM-g-MS) as the activator. Both the anhydride and the ester groups were found to be effective in the activation of CL polymerization, most likely through the fast formation of N-carbamoylcaprolactam groups:

$$(1)$$

$$(I)$$

The macromolecular activator I thus formed should play a role analogous to that of conventional activators such as N-acetylcaprolactam (AcCL). The following polymerization scheme is therefore proposed:

where equilibria (2), (3) and (4), (5) represent, respectively, the initiation and the propagation steps. The polymerizations were carried out in bulk or in the presence of a diluent by adding solutions of NaCL in CL either to dispersion of functionalized EPM in CL

or to xylene solutions of CL and EPM-g-MS. Also investigated was
the influence of adding AcCL on the grafting degree. In every case
the polymerization mixture became grossly heterogeneous within a few
seconds. Some typical results are summarized in Table 2. It can be
observed that high NaCL/CL molar ratios were used as a consequence
of $CL^{\ominus}$ consumption in the formation of the macromolecular initiator
(I). In fact, when equimolecular amounts of $CL^{\ominus}$ and ester groups of
EPM-g-MS were employed, the polymerization of CL did not occur
because of the absence of $CL^{\ominus}$ required in the initiation and
propagation equilibria (2)-(5).

Runs 1-4 clearly show that, in the absence of AcCL, even if the
CL conversion does not exceed 20%, graft copolymers having up to
32% by weight of polyamide can be obtained. It is interesting that,
except for run 4, the polymerized CL was found mostly as grafted
polyamide chains and, to a smaller extent, as PA 6 homopolymer.
Slightly better results were obtained in the bulk polymerizations
with respect to those carried out in the presence of xylene under
the same experimental conditions.

A quite different picture arises when AcCL is added to the
polymerization mixture (runs 5-7). First, a sharp increase of CL
conversion is observed. Secondly, not only does the amount of PA 6
homopolymer reach high values, as is to be expected, but the grafting
degree also increases with respect to polymerizations performed in
the absence of AcCL. These results are quite in keeping with those
reported by Braun et al.[3] for the synthesis of polyethylene-g-SA
copolymers and they can be explained by assuming that polymeric
amidoanions undergo an acylation reaction by acyllactam end groups
of growing chains:

A ring opening on CL end groups of polyamide side chains, (6), or
a $CL^{\ominus}$ release, (7), will occur according to the type of carbonyl
group involved in this reaction. By this route polymeric amidoanions
initiated from AcCL may contribute to an additional growth of polyamidic side chains grafted on EPM. Secondary reactions of the anionic
process may also play a similar role. In fact, it is well known[8]
that a strongly basic reaction medium promotes proton extraction

Table 2.   Anionic Polymerization of ε-Caprolactam in the presence of EPM-g-MS

Run	1	2	3	4	5	6	7
Diluent	xylene	xylene	—	—	xylene	xylene	—
T (°C)	139	180	180	210[1]	139	139	139
Wt.% MS in EPM-g-MS	6.5	6.5	6.5	2.00	6.5	6.5	6.5
Molar ratio MS/CL	1/100	1/100	1/100	1/100	1/100	1/200	1/200
Molar ratio NaCL/CL	6/100	3/100	3/100	3/100	3/100	3/100	3/100
Molar ratio AcCL/CL	—	—	—	—	1/100	1/200	1/200
CL conversion (%)	10	13	20	16	73	64	83
Wt.% of PA 6 in EPM-g-PA 6[2]	18[3]	25	32[3]	6	46	33	36
PA 6 Yield (wt.%)	5	3	9	12	58	58	77
η inh of PA 6[4]	—	0.30	—	—	0.60	0.90	1.2

1) Polymerization carried in a sealed vial,
2) Determined by elemental analysis except runs 1,3.
3) Determined by gravimetric analysis.
4) In m-cresol at 25° C   (c = 0.5 dl/g).

at the carbon atoms in α position to the imide group, thus forming carbanions of the following structure:

These carbanions can condensate with acyllactam end groups of grafted chains, i.e.:

(8)

(9)

Again, these equilibria may be, at least in part, responsible for the increased grafting degree in the presence of AcCL.

The EPM-g-PA 6 copolymers were isolated from the polymerization mixture as residue after repeated methanol extractions to remove CL and CL oligomers and formic acid extractions to remove PA 6 homopolymer from the crude polymerization product. The absence of ungrafted EPM was shown by xylene extraction of the purified copolymers, which gave practically no extraction products. The copolymers were characterized by elemental analysis, IR spectroscopy and DSC investigations. They are slightly crosslinked as indicated by their insolubility and by the fact that they do not flow in the melt. The crosslinking phenomenon has been found in other PA 6 graft copolymers prepared by the anionic route (5). This can be easily explained on the basis of the above reported equilibria (6)-(9). In fact, if both polyamide chains involved in the quilibria are linked to only one or to two different polyolefinic backbones, formation of intra- or inter-molecular polyamide bridges, respectively, will result. Acid hydrolysis of the graft copolymers yielded a soluble polymeric product identified by Ir analysis as EPM-g-succinic acid, thus confirming that the crosslinks do not involve the EPM chains directly.

In DSC thermograms EPM-g-PA 6 copolymers show a first order transition in the 205°-210° C temperature range, attributable to the melting of a crystalline polyamidic phase (T_m of PA 6 $\simeq 220^{\circ}$ C). The mass crystallinity values, as determined from the apparent melting enthalpies and corrected for the copolymer composition, are significantly lower (20-25%) than those generally found by us in the recovered PA 6 ($\geq$35%). These results suggest that the average

chain length of the PA 6 grafts are high enough to permit ordering
of the side chains in a crystalline phase in spite of the strong
conformational constraints due to the copolymer structure.

CONCLUSIONS

The incorporation of anhydride functional groups on amorphous
EPM copolymers can be accomplished by solution grafting of MAH molecules onto EPM by means of a radical initiator (BPO). Parameters
such as reaction time and BPO and MAH concentrations influence the
grafting degree. Investigations on the reactivity of the grafted
anhydride groups suggest that EPM/PA 6 graft copolymers should be
formed during the preparation of EPM/PA 6 blends by melt mixing PA 6,
EPM-g-SA and EPM. The graft copolymer may act as interfacial agent
in these blends. Formation of EPM/PA 6 graft copolymers has been
shown to occur when CL is anionically polymerized in the presence
of functionalized EPM. Such a preformed graft copolymer could
actually be unsuitable to act as interfacial agent owing to the
crosslinks between polyamide chains.

ACKNOWLEDGMENTS

Authors gratefully acknowledge the financial support of
"Progetto Finalizzato Chimica Fine e Secondaria del C.N.R."

REFERENCES

1. D. R. Paul, "Interfacial Agents for Polymer Blends", in: "Polymer
    Blends", D.R. Paul and S. Newman, eds., Academic Press,
    New York (1978).
2. F. Ide and A. Hasegawa, J.Appl.Polym.Sci., 18:963 (1974).
3. D. Braun and U. Eisenlhor, Angew.Makromol.Chem., 55:43 (1976).
4. M. Matzner, D.L. Shober and J.E. McGrath, Europ.Polym.J., 9:469
    (1973).
5. M. Matzner, D.L. Shober, R.N. Johnson, L.M. Robeson and
    J.E. McGrath, in: "Permeability of Plastic Films and Coatings",
    H.B. Hopfenberg, ed., Plenum Press, New York (1974).
6. M. Avella, R. Greco, N. Lanzetta, G. Maglio, M. Malinconico,
    E. Martuscelli, R. Palumbo and G. Ragosta, in: "Polymer Blends:
    Processing, Morphology and Properties", E. Martuscelli, R
    R. Palumbo and M. Kryszewski, eds., Plenum Press, New York
    (1980).
7. D. Braun and U. Eisenlhor, Angew.Makromol.Chem., 58/59:227
    (1977).
8. J. Sebenda, J.Macromol.Sci., Chem., 6:1145 (1972).
9. P. Biernacki and M. Włodarczyk, Eur.Polym.J., 16:843 (1980).

A MECHANICAL STUDY OF POLYETHYLENE-GLASS BEADS ADHESION

M. Pegoraro, A. Penati, E. Cammarata, and M. Aliverti

Institute of Industrial Chemistry
Polytechnic of Milan
Italy

INTRODUCTION

In order to obtain useful composite materials compatibility of glass with polyolefins is required. Compatibility can be attained by chemical modifying the polyolefins or the glass surface or both. We tried as a first approach[1] the grafting of a polar monomer, such as acrylic acid on polypropylene, but we found that the formation of small domains of polyacrylic acid gives rise to a small interaction surface with the glass. In this paper we present the results obtained when modifying the glass surface with short paraffinic chains and using polyethylene as the matrix of the composite. Stearylisocyanate (SIC) and isopropyltriisostearyltitanate (TTS) were used as modifiers of type A glass, enriched in the oxidryl content. The reaction outline is given in Fig. 1 but side reactions were also found to occur. The structure of the paraffinic chains fixed on the glass is practically linear in the SIC case, and highly branched in the TTS case.

Polyethylene (PE) was a low density one (0.92 g/ml). Mechanical tests showed that the best reinforcing results are reached with SIC treatment.

In this paper, detailed evidence is given concerning the fundamental aspects of mechanical tests as a means of evaluating interphase adhesion.

A new discussion is presented about the tensile work-adhesion relationships.

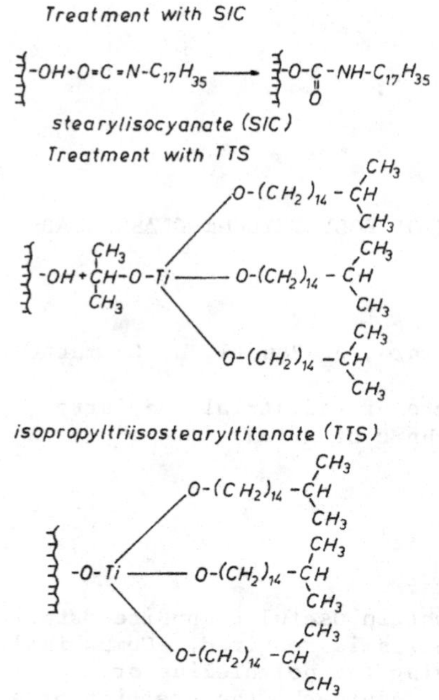

Treatment with SIC

$$\text{≡-OH+O=C=N-C}_{17}\text{H}_{35} \longrightarrow \text{≡-O-}\underset{\underset{O}{\|}}{C}\text{-NH-C}_{17}\text{H}_{35}$$

stearylisocyanate (SIC)

Treatment with TTS

isopropyltriisostearyltitanate (TTS)

Fig. 1.   Reaction scheme of hydroxyl glass surface groups with SIC
           and TTS.

EXPERIMENTAL

Materials used

- Low density polyethylene ZF-2000 Montedison:
  density: 0.920 g/ml; melt flow index (ASTM D1238: 0.25 g/10');
- Glass beads 3000 CP00 of Tradex Company glass composition:
  SiO_2 = 72%, Na_2O = 14%, CaO = 8%, MgO = 4%, Al_2O_3 = 2%
  Beads diameter distribution between 4 and 44 µm;
- Stearylisocyanate (SIC) - a commercial product of Bayer CO.,
  a mixture of $C_{18}H_{37}NCO$ and $C_{16}H_{33}NCO$;
- Isopropyltriisostearyltitanate (TTS) - a commercial product
  of Kenrich Petrochemicals.

Preparation of the composites

     Glass beads were hydrolized with distilled water at 150° C in an
autoclave for 24 hours. After washing with water and very accurate

drying, the OH content of the glass surface was analyzed using LiAlH$_4$ according to the method suggested by Krynitsky[2]. The amount of OH was found to increase from 0.007 me/g to 0.02 ÷ 0.03 me/g. Dry glass beads were immersed in liquid (SIC or TTS) reagents, filtered and washed with acetone for SIC, and with n-hexane for TTS treatment. Carbon and hydrogen elementary analysis was carried out by combustion to determine the reaction yield. The modified beads used for the mechanical tests met the following conditions: SIC treated: C = 3.07%, H = 1.07%; TTS treated: C = 1.16%, H = 0.76%. Treated beads were gradually added to polyethylene melt sheet by calendering at temperatures in the range of 125–135° C. The time of calendering was about 15 minutes.

The different blends were successively molded by compression at 130° C with a pressure of 50 kg/cm^2 applied for 5 minutes to obtain sheets of appropriate thickness (1 ÷ 6 mm). The samples for the different mechanical tests were obtained by cutting or punching. The tests were made by an Instron dynamometer at 23° C, 50% relative humidity, according to ASTM D 638 for the tensile test (elongation speed 1 cm/min) and to ASTM D 790 for the flexural tests (speed 0.2 cm/min). The main results are presented in Figs. 2, 3, 4, 5, 6, 7, 8 where the mean values are given; standard deviation was in the range ±0.3% ÷ ±8.5%. The complex modulus (mechanical dynamical tests) was measured using an I.N.U.A. resonance apparatus of Bordoni[3].

RESULTS AND DISCUSSION

Adhesion can only be evaluated phenomenologically by mechanical tests. Two classes of properties can be considered: the mechanically reversible ones, as the elastic modulus (real part), and the irreversible ones, as the yield and the ultimate properties.

The modulus can be predicted by the elastic theory under the assumption of equal strain of the two phases (good adhesion), in the range of small strains. Adhesion is related[4] to the radial pressure exerted on the beads due to the relative shrinkage of the polymer and the filler having different thermal expansion coefficients. Adhesion is therefore guaranteed in the case of small deformations and the influence of the chemical interaction on the interphase elastic modulus cannot be easily demonstrated.

On the other hand, ultimate properties are largely used to evaluate the adhesion and the chemical treatment efficiency but, again, the relevant theories are in an elementary stage: they often consider the inclusion of one filler only and assume isostrain behaviour of the two phases and a σ, ε linear behaviour until the ultimate properties are reached[5,6] in the case of brittle composite materials. In the case of thermoplastics, the yield point appears to substitute adequately, with the same approximation, the break point.

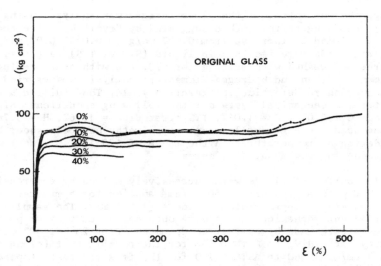

Fig. 2.   Stress strain tensile curves of composite PE-glass untreated
          beads at different volume fraction $\emptyset$.

Fig. 2 shows typical stress-strain tensile curves of the composites PE-untreated beads, characterized by different untreated glass
volume fractions $\emptyset$. The polyethylene behaviour is also shown ($\emptyset = 0$).
The initial part of the curves, in a first rough approximation, can
be considered linear. The yield is always present. Beyond the yield
point plastic behaviour is observed: the tensile strength and strain
decrease with increasing $\emptyset$.

Similar results were obtained for the compsites of polyethylene
and SIC and TTS treated beads: the quantitative results will be
discussed below.

Yield point

A limit value for the yield strain ε_y can be obtained using
the simple series model of Nielsen[7]. According to this model, perfect adhesion is assumed, and the actual strain of the polymer (ε_m)
is higher than the strain of the composite (ε_c) due to the fact
that glass does not elongate; the relationship is

$$\frac{\varepsilon_c}{\varepsilon_m} = 1 - \emptyset^{1/3} \tag{1}$$

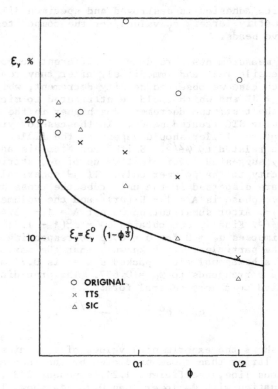

Fig. 3. Yield strains ε_y of PE-glass beads composites against ϕ. Different bead treatments are indicated with different marks.

If one assumes that the polymer yields at the same strain in the filled system as the bulk pure polymer does, the yield strain of the composite should be

$$\varepsilon_y = \varepsilon_y^0 \, (1 - \phi^{1/3}) \, , \qquad (2)$$

where ε_y^0 is the yield strain of the unfilled polymer.

The experimental values of yield strain, ε_y, are shown in Fig. 3. The experimentaly measured strains are higher than the theoretically predicted ones and increase in the following order: SIC, TTS, and no treatment. This effect should be attributed to some non-homogeneous deformation mechanism. The yielding phenomenon in ductile filled polymers is often attributed[8] to a crazing effect or to a de-wetting effect in which adhesion is destroyed and specimen dilation occurs; this could justify larger ε_y values for the composites containing fewer adhesive beads.

Density measurements were done on different composite samples before the tensile test and immediately after they reached the yield point. In each case we observed density decrement, which was of the order of $0.5 \div 1\%$ and which could be attributed to microvoid formation during the test: the decrement was higher in the case of composites containing SIC treated beads. In the case of yield stress σ_y, the effects of the filler should depend on the filler surface area and therefore related to $\phi^{2/3}$. Smith[9] and Nicolais and Narkis[10] theoretically suggested lower limit value of σ_y, attributing the loading capacity to the polymer only. If n^3 spherical particles of radius r are dispersed in the unit cube the cross section of the continuous phase is $A = 1 - \Pi \cdot (nr)^2$ and the volume fraction is $\phi = \frac{4}{3}\Pi(nr)^3$. After substitution we get $A = 1 - (3/4\Pi)^{2/3}\phi^{2/3} = = 1 - 1.21\ \phi^{2/3}$. Finally one obtains $\sigma_y = \sigma_y^o(1 - 1.21\ \phi^{2/3})$. The factor 1.21 imposes $\sigma_y = 0$ for $\phi = 0.75$ corresponding to direct contact of the particles: it is known[11] that the maximum packing factor ϕ_m in a hexagonal close packed system is 0.74 and that random close packing corresponds to $\phi_m = 0.637$. The preceding equation can be modified to a more general form:

$$\sigma_y = \sigma_y^o \left[1 - (\frac{\phi}{\phi_m})^{2/3}\right] . \tag{3}$$

Fig. 4 shows the experimental values of σ_y agains ϕ, which are in each case larger than those predicted by the theoretical equation (see continuous line, coefficient 1.21, exponent 2/3). Using the preceding equation with ϕ_m lower than 0.74, σ_y shows lower values again. The relatively high values of σ_y in comparison with the theoretically predicted ones must be attributed to adhesion phenomena which appear to be always present and increase in the order: untreated; TTS and SIC treated fillers. Thermal shrinkage justifies the existence of some adhesion even in the case of untreated beads.

Some comment is necessary on the behaviour of the experimental function σ_y (ϕ) which generally shows a decrement with increasing ϕ, with the exception of the SIC treated beads in the low ϕ range. According to some authors[12,13], in the case of perfect adhesion, and for certain fillers and glassy polymer systems which do not show yield, there is an upper limit of the strength of the composite which simply equals the strength of the polymer itself. According to other authors[6], the principle of load partition between the two phases should be accepted. The simplest form of this principle can be written as $\sigma = \bar{\sigma}\phi + \sigma_m(1-\phi)$, where $\bar{\sigma}$ is the mean stress acting on the beads (depending on adhesion) and σ_m is the stress acting on the polymer when the filler is present and takes into consideration the stress concentration factor[14].

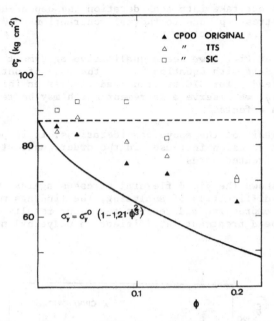

Fig. 4. Yield stress σ_y of PE-glass beads composites against $\emptyset$. Different bead treatments are indicated with different marks.

This equation is accepted as true, until the ultimate strength of the composite is reached and appears to be in agreement with the experiments when applied at break in the case of glass beads and polyesters[6]. The preceding equation allows for higher strength of the composite than of the matrix. Also Nielsen[7] accepts the load partition principle and the possibility of obtaining particular fillers composites endowed with higher strength than the polymer itself. The preceding equation can be rearranged as:

$$\sigma = \sigma_m + (\bar{\sigma} - \sigma_m)\,\emptyset \,, \qquad (4)$$

and used in our case (ductile matrix), as said above, only until the yield point is reached. Low and mean ahhesion corresponds to $\sigma \ll \sigma_m$ and $\bar{\sigma} < \sigma_m$; good adhesion to $\bar{\sigma} > \sigma_m$.

The equation is a linear one if parameters (as a first rough approximation) are considered constant with $\emptyset$. The equation does not take into account the agglomeration effects which are an increasing function of $\emptyset$. Agglomerates are weak points in the materials and act as strong stress concentrators[15]. The linear equation does not take into consideration the dependence of the compressive stress p due to thermal contraction on $\emptyset$[21,22]. As is known, $\bar{\sigma}$ is a function of p[4].

Looking at Fig. 4 we see a qualitative agreement of the experimental data with equation (4) in the case of untreated and TTS treated beads; for SIC treated beads, after an initial small increase of σ_y, we observe a decrement which may be related to agglomeration effects.

On the basis of the above considerations, it is possible to conclude that adhesion increases in the order: untreated, TTS treated, SIC treated beads.

Fig. 5 shows the yield flexural stresses against $\emptyset$. Notwithstanding the difficulties of analyzing, the findings made possible by this new compressive and tensile test, our results confirm that SIC and TTS bead treatment is efficient in polyethylene reinforcement.

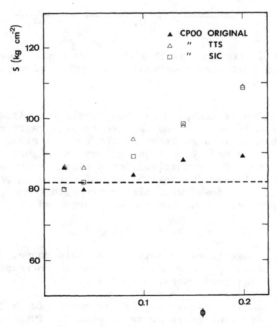

Fig. 5.   Yield flexural stress of different composites against $\emptyset$. The different treatments are indicated with different marks.

Ultimate properties

Tensile strength (Fig. 6), and strain (Fig. 7), continuously decrease with the increase in $\emptyset$. This is a well known effect: premature fracture occurs due to the amount of defects (flaws) which is directly related to the filler fraction.

SIC and TTS adhesives treated beads, show lower σ_R and ε_R than untreated beads, as if good adhesion could reduce the ease of plastics flow and alignment of the chains.

Fig. 8 shows the characteristic appearance of beads treated with SIC observed in the fracture surface. Adhesion is evident; on the contrary, untreated beads appear to be bald.

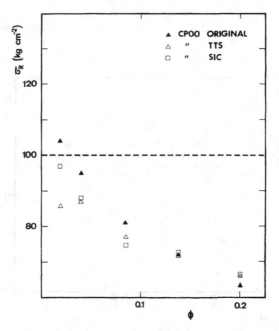

Fig. 6.   Tensile strength of the different types of composites.
          See different marks.

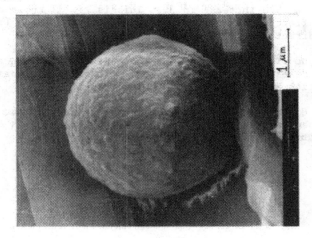

Fig. 8. Beads treated with SIC observed in a fracture surface.

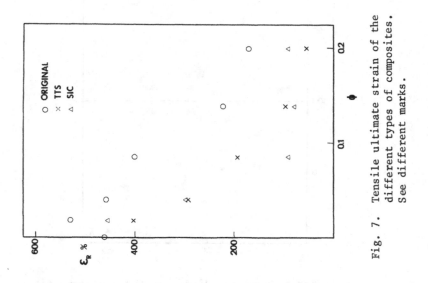

Fig. 7. Tensile ultimate strain of the different types of composites. See different marks.

Modulus of elasticity

Many theories have been advanced for predicting the modulus of filled composites. The Kerner[16] theory is often used for the G modulus in the case of filled systems containing spheres; Halpin--Tsai[17] modified the Kerner equation in a more general form; Lewis and Nielsen[18] suggested a further modification by taking into consideration the packing factor and obtained, in the case of E modulus, the following equation:

$$\frac{E}{E_m} = \frac{1 + AB\phi}{1 - B\phi\psi} \tag{5}$$

which is limited to isotropic materials when $G/G_m \simeq E/E_m$. E and E_m in the above equation are the Young's elasticity moduli of the composite and of the matrix, G and G_m are the corresponding shear moduli, A depends upon the Poisson ratio of the matrix ν_m and the geometry of the filler: $A = (7-5\nu_m)/(8-10\nu_m)$, B is a function of the ratio of the filler modulus E_g and the matrix modulus E_m: $B = (E_g/E_m-1)/(E_g/E_m + A)$, and ψ is the packing factor: $\psi = 1 + [(1+\phi_m)/\phi_m^2]\phi$ which is an empirical function of the maximum packing factor of the filler ϕ_m.

Fig. 9 shows the flexural Young modulus of our different composites; Fig. 10 shows the mechanical dynamical flexural modulus measured at room temperature in the range of 500 c.p.s. in the case of untreated and SIC treated beads. The theoretical curves (continuous lines) were obtained with the parameters $\nu_m = 0.38$, $E_g = 700\ 000\ kg/cm^2$, $E_m = 6\ 700\ kg/cm^2$ in the case of the dynamical tests, and $E_m = 2\ 150\ kg/cm^2$ in the case of the flexural tests, $\phi_m = 0.60$ random loose packing for α line and $\phi_m = 0.74$ for β line.

The theories considered above, as has already been indicated cannot be easily applied to give an explanation of the differences among the experimental results related to the different treatments. For a given matrix, the Eq. (5) can account for the differences only by changes of ϕ_m and of E_g/E_m. The variation of ϕ_m from 0.60 to 0.74 (maximum excursion range) cannot give in our case variations of E/E_m larger than a few pertcent. Variation of E_g/E_m, on the contrary, has a great influence on E/E_m and could correspond to the change of the treated filler nominal modulus E_g, depending on the treatment. Partial adhesion should correspond to a lower modulus. According to this hypothesis, adhesion should increase in the order: untreated and SIC treated beads.

A three-phase model could probably correspond more closely to reality[19,20].

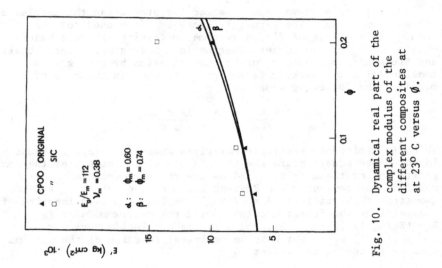

Fig. 10. Dynamical real part of the complex modulus of the different composites at at 23° C versus ∅.

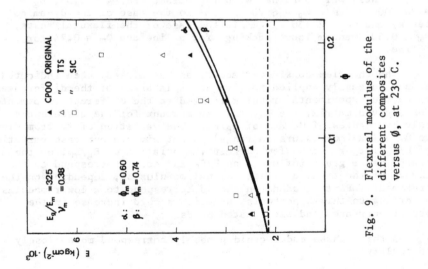

Fig. 9. Flexural modulus of the different composites versus ∅, at 23° C.

Mechanical work

It is well known that the work necessary to obtain a given strain of a sample of unit volume subjected to a tensile experiment is given by the area under the stress-strain curve. Considering a prismatic sample of length 1, volume V, surface S under the force f, the first thermodynamics principle allows us to write the differential equation

$$fd1 = dU-dQ + pdV + \gamma dS \ , \qquad (6)$$

where dU is the sample internal energy variation, -dQ is the heat given out to the environment, pdV is the expansion work against the external pressure p, and γdS is the work needed to increase the sample surface under the surface tension; this expression can be important in the case of crazes. Dividing each term by $A_0 L_0$ (initial volume of the sample) and integrating between two given strain values at p and T constant, one obtains:

$$\frac{W}{V} = W' = \int_{\varepsilon_1}^{\varepsilon_2} \sigma d\varepsilon = \Delta U' - \Delta Q' + p\Delta V' + \gamma \Delta S' = \Sigma \Delta I' , \qquad (7)$$

where the prime is related to the sample unit volume, and where I has the meaning of U, Q, pV, γS. Each term can be related to the different components of the composite (matrix m and glass g); due to the importance of the surface effects, it is useful to consider them separately.

Therefore, the equation (7) can be written as

$$W' = \Sigma_m \Delta I' + \Sigma_g \Delta I' + \Sigma_i \Delta I' . \qquad (8)$$

The terms related to the matrix and to the filler can be considered proportional to their fractions 1-$\emptyset$ and $\emptyset$. On the contrary a non-linear dependence upon $\emptyset$ is to be expected for the interphase surface effects. The interphase surface of a system of n^3 spheres of radius r per unit volume is linearly correlated to $\emptyset$ ($S_i = 3\emptyset/r$), but many important phenomena, such as composite volume and surface increase during the loading (for example due to the crazing) largely depend, in a way that is not fully understood yet, on the formation and growth of defects starting from certain points of the filler surface without involving the entire surface. In addition to that, with increasing $\emptyset$, particle agglomeration is favoured and acts as if the stress concentration factor should increase[8]. Therefore the mechanical and energetic surface induced effects are a complicated function of $\emptyset$.

For these reasons we can write the equation 8 in the form

$$W' = A(1 - \emptyset) + B\emptyset + \psi(\emptyset) \quad . \tag{9}$$

Factor A applies to the matrix only; factor B, which is characteristic for the filler, is practically very small in the case of a rigid filler and can be neglected. The function $\psi(\emptyset)$ gives the energetic term due to processes which take place not only on the surface but also in the matrix due to the matrix-surface interaction.

The equation can be studied in any given interval of ε, separating, for example, the initial stress-strain zone (where ΔU is high) and the zone beyond the yield point where, for example, the ΔU terms in the case of crystalline polymers are presumably small.

Fig. 11 shows the experimental results obtained following integration of the σ, ε curves in the range $\varepsilon_1 = 0$; $\varepsilon_2 = 0.50$.

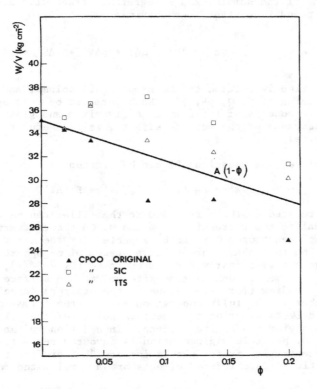

Fig. 11.   Specific work of the different composites against $\emptyset$. Points correspond to the integration limits $\varepsilon_1 = 0$, $\varepsilon_2 = 0.5$.

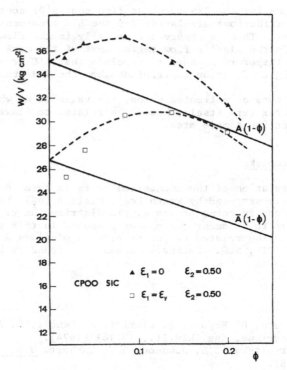

Fig. 12.   Specific work of SIC composites against ϕ
▲ in the region $\varepsilon_1 = 0$; $\varepsilon_2 = 0.50$
□ in the region $\varepsilon_1 = \varepsilon_y$; $\varepsilon_2 = 0.50$.

This upper limit was chosen as a common reference for all the composites, because it is lower than the least strain at break (ε_R) of all the composites.

We observe that the composite toughness $W' = W'/V$ can be increased, relative to the matrix toughness, by the addition of fillers when they are treated with SIC and TTS. In the case of untreated beads, on the other hand, composite toughness always decreases with increasing ϕ and is lower than matrix toughness.

In Fig. 12, W' represents the SIC composite only. The upper dotted curve illustrates the case of Fig. 11 with the integration limits $\varepsilon_1 = 0$; $\varepsilon_2 = 0.50$. The lower dotted line shows W' in the range $\varepsilon_1 = \varepsilon_y$, $\varepsilon_2 = 0.50$. The matrix terms are indicated respectively from the linear equations $A(1-\phi)$ and $\overline{A}(1-\phi)$. In both cases, a strong influence on the surface effects is observed. Comparing

similar graphs for the TTS case, we find that $\psi(\emptyset)$ contributes to
W' which contributions are larger for the SIC treatment than for the
TTS treatment. This is observed especially in the flow region
$(\varepsilon > \varepsilon_y)$. In the plastic flow region most of the work is dissipated
as heat and therefore we have to conclude that SIC treated beads
present a higher friction coefficient than the TTS treated ones.

In the case of untreated beads, the value of W' which is lower
than the matrix term itself, has to be related to flaws which reduce
the cross-section sample area.

Concluding remarks

Interpretation of the mechanical tests in terms of the sim-
plified models proposed by Smith (ε_y), Nielsen (σ_y), Lewis and
Nielsen (E) and in terms of the simple distribution of the energetic
contributions to W' among the phases proposed in this paper, suggests
that the adhesion related to our treatments of glass A beads increases
in the order TTS, SIC. Untreated beads were found to have low
adhesion.

REFERENCES

1. M. Pegoraro, G. Pagani, P. Clerici, A. Penati, G. Alessandrini,
    R. Maggione, Ing.Chim.It., 10:161 (1974).
2. J. A. Krynitsky, J.E. Johnson, H.W. Carhart, J.A.C.S., 70:486
    (1948).
3. P. G. Bordoni, Il Nuovo Cimento, 4:177 (1947).
4. J. O. Outwater, Modern Plastics, 33:156 (1956).
5. A. Kelly, Strong Solids, Clarendon Press, Oxford (1973).
6. J. Leidner, R.T. Woodhams, J.Appl.Polym.Sci., 18:1639 (1974).
7. L. E. Nielsen, J.Appl.Polym.Sci., 10:97 (1966).
8. L. E. Nielsen, Mechanical Properties of Polymers and Composites,
    M. Dekker, New York (1974), vol. 2, p. 408.
9. T. L. Smith, Trans. of the Soc. of Rheology TSRHA, 3:113 (1959).
10. L. Nicolais, M. Narkis, Polym.Eng. and Sci., 11:195 (1971).
11. R. K. McGeary, J.Am.Ceram.Soc., 44:513 (1961).
12. A. S. Kenyon, H.J. Duffey, Polym.Eng. and Sci., 7:189 (1967).
13. L. Nicolais, L. Nicodemo, Polym.Eng. and Sci., 13:469 (1973).
14. H. Hajo, W.T. Toyashima, 31st ANTEC, SPE, Montreal Canada
    (1973) p. 163.
15. L. E. Nielsen, see ref. 8, p. 413.
16. E. H. Kerner, Proc.Phys.Soc., 69B:808 (1956).
17. J. Halpin, J. Kardos, Polym.Eng. and Sci., 16:403 (1976).
18. L. E. Nielsen, J.Appl.Phys., 41:4626 (1970).
19. G. C. Papanicolau, P.S. Theocaris, G.D. Spathis, Colloid
    and Polym.Sci., 258:1231 (1980).

20. H. H. Kausch, "Polymer Fracture", Springer, Berlin (1978).
21. F. Laszlo, J.Iron Steel Inst., 147:193 (1943).
22. L. Nicolais, R.A. Mashelkar, Int.J.Polym.Mat., 5:317 (1977).

THE STRUCTURE AND TENSILE PROPERTIES OF COLD

DRAWN MODIFIED CHALK FILLED POLYPROPYLENE

T. Kowalewski, A. Gałęski and M. Kryszewski

Centre of Molecular and Macromolecular Studies
Polish Academy of Sciences
90-362 Łódź, Poland

INTRODUCTION

The aim of the present work was to investigate the correlation
of structure and tensile properties of highly oriented-highly filled
isotactic polypropylene. On the basis of previous studies it is
assumed that the obtained results could be generalized for other
polyolefins and, possibly, other semicrystalline polymers. As in
all further considerations the concept of system heterogeneity plays
an important role, we shall start with some remarks on the types
of heterogeneity in polymer-filler systems. Following that the
structure and tensile properties of the system are deduced with the
help of a semiempirical model proposed on the basis of some results
described in the previous works on this subject. A set of experimental procedures including deformation at constant elongation rate,
TEM, X-ray diffraction will be used to test the theoretical
predictions.

Heterogeneities in polymer-filler systems

The heterogeneity of any system depends on two factors:
- relative difference in the properties of the elements constitut-
ing the considered system,
- strength of the interactions between these elements.
The system is relatively homogeneous when both factors are low, and
becomes increasingly more heterogeneous as they increase.

In the case of a system containing semicrystalline or elastomeric polymer and mineral filler the first of the above mentioned

factors implies its heterogeneity which is enhanced by strong polymer-filler   adhesion. The effective length of the polymer matrix in
the deformation direction is then decreased due to the presence of
strongly bound filler particles. Stress and strain concentrate at
the pole sides of the filler particles. Consequently – the sample
modulus and strength increase. This effect is widely utilized in the
case of filler reinforced elastomers[5,6]. In this case the filler
plays an active – reinforcing role and, as a result, the shape of
the filled elastomers is maintained.

    Fillers are introduced into semicrystalline polymers for various
reasons. The filler is supposed to play a rather passive role
lowering the polymer volume in the new material. The mechanical
properties of such a system ought not to depart too much from the
pure polymer i.e. its elasticity, plasticity, susceptibility to
plastic deformation and impact strength should all be preserved.
It might be supposed that in this case the tendency towards system
homogeneity should be preferred. It was shown[1-3] that introduction
of liquid (oligomer of ethylene oxide – OEO) at polymer-filler
interface (reduction of the strength of the interactions between
the elements of the system constitute a step towards homogeneity)
does produce desired results, in that is makes possible e.g. cold
drawing of highly filled samples to high draw ratios (λ = 5 ÷ 6).

    In one of the previous works discussing the role of the liquid
interface agent[4] it was found that during deformation of the sample
of i-PP filled with OEO modified chalk its volume increased with the
draw ratio (up to 1.5 – 2 times for 40 wt.% content of chalk), which
can be taken to indicate that voids form around the filler particles.
The heterogeneity of the polymer matrix deformation resulting from
it is of the opposite character in comparison with the case of strong
adhesion. The filler particles reduce the sample cross-section in the
plane normal to the deformation direction. As a consequence stress
and strain concentrate at the equatorial sides of the grains. It may
be that some fraction of the polymer matrix located at the pole sides
of the grains remains less strained or is not strained at all.
Thus, during the deformation of a polymer filled with modified chalk,
a new interesting heterogeneous system is obtained containing voids,
nonuniformly oriented polymer matrix and practically inert filler
particles. Such a system combines the properties of porous, oriented
and filled materials which makes it interesting from both the
theoretical and practical point of view.

    The unbound filler particles cause nonuniform external stress
distribution inside the samples resulting in nonuniform molecular
orientation. There are certainly regions of very high molecular
orientation and of very low orientation. The scale and position of
these disturbances are one of the subjects of the present work.
The next section will be devoted to evaluation of a phenomenological

model describing the origin of such a structure and its relation to
the tensile properties.

THEORETICAL PREDICTIONS

In all further considerations we treat the filled polymer sample
macroscopically, using a scale large in comparison with the size of
its largest structure heterogeneities (i.e. filler particles, voids
around them etc.). The following assumptions are also made:
A1 Minimum or no adhesion between the polymer and filler is present.
    Only the polymer matrix plays an active role during the deforma-
    tion. Microscopic parameters which are averages (draw ratio,
    mean direction of orientation, orientation distribution function)
    are taken over the volume occupied by the polymer matrix.
A2 The polymer matrix preserves its volume during deformation.
    (Poisson ratio equal to 0.5).
A3 Filler particles are distributed uniformly in the matrix.
A4 In macroscopical terms, the considered volume of the sample
    deforms uniformly.
The molecular orientation of the polymer matrix is described by
means of a vector field $\vec{n}(x,y,z)$, with $\vec{n}$ being the unit vector
(director) oriented along the mean molecular orientation direction.
The molecular orientation of uniaxially drawn homogeneous polymers
might be approximated, if one neglects the influence of the outer
shape of the sample, to the cylindrical symmetry described by the
orientation distribution function of the form:

$$f(\gamma) = a(\gamma) \sin \gamma \qquad\qquad (1)$$

where γ is the angle between director and orientation directions,
and $_0\int^{\Pi/2} f(\gamma)d\gamma = 1$. Use will be made of the well known orientation parameter

$$S = \tfrac{1}{2}(3 < \cos^2\gamma > -1) , \qquad\qquad (2)$$

where $< \cos^2\gamma > = {_0}\int^{\Pi/2}\cos^2\gamma\ f(\gamma)d\gamma$. In perfectly uniaxially
oriented samples $a(\gamma) = \delta(\gamma)$, where $\delta(\gamma)$ - Dirac's delta function
and $S = 1$, while for totally unoriented sample $a(\gamma) = 1$ and $S = 0$.
In the orientation process $f(\gamma)$ is also a function of the draw ratio
λ, $f(\gamma, \lambda)$. Since the inert filler particles introduce the external
stress nonhomogeneity and, by the same token, the molecular orientation nonhomogeneity of the matrix, we assume that the system under
consideration is a "mixture" of unoriented and oriented volume
fractions changing with the draw ratio and equal to $k(\lambda)$ and
$1 - k(\lambda)$ respectively. The mean draw ratio in the strained sample
can be defined as (see A1, A4)

$$\lambda = \left[1 - k(\lambda)\right] \lambda' + k(\lambda) \qquad\qquad (3)$$

where λ' denotes the draw ratio in the strained fraction of the polymer matrix.

The orientation distribution function for such a two phase system will be obtained by superposition of the distribution functions of both phases (see A1, A4)

$$f_{fs}(\gamma,\lambda) = k(\lambda) \sin \gamma + \left[1 - k(\lambda)\right] f(\gamma,\lambda') \quad . \tag{4}$$

In order to find the values of $k(\lambda)$ we will modify the model giving the form of $f(\gamma,\lambda)$ for pure polymer proposed by Kratky (originally[7], here quoted after[8], see also[9],[10]). This model termed "pseudo affine" for the reasons discussed in the foot-note, has been shown to give e.g. a qualitative fit to experimentally found birefringences vs. draw ratio for polyethylene, nylon, polyethylene terephtalate and polypropylene[9],[10]. In spite of the fact that the model neglects the structural changes taking place during the deformation of semicrystalline polymer, it seems to be of a great value due to its ability to give qualitative explanation of orientation dependent phenomena[10]. Let us consider some cylindrical volume element of the sample with initial dimensions r_o (radius), h_o (height - parallel to the deformation direction coinciding with the OZ axis of the Cartesian coordinate system). In undeformed state director makes with the deformation direction some angle γ_o (see Fig. 1a). After uniaxial deformation to the draw ratio $\lambda = h/h_o$ the director is aligned at some

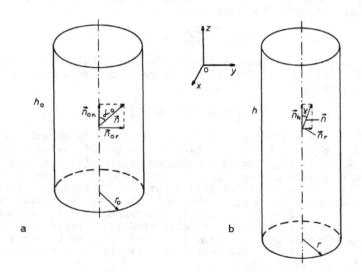

Fig. 1.   Affine deformation scheme.   Cylindrical volume element
          of a sample before   (a) and after   (b) deformation.

new angle γ and the volume of this element increases due to the formation of voids around non-adhering filler particles

$$V = \Pi \, r^2 \, h = B(\lambda) \, \Pi \, r_o^2 \, h_o = B(\lambda) \, V_o \quad . \tag{5}$$

We will assume that the molecular orientation follows macroscopic deformation i.e. the lengths of projections of director at the deformation direction (n_{oh} - before deformation, n_h - after deformation) and at OXY (cylinder base) plane (n_{or} - before deformation, n_r - - after deformation) (see Fig. 1b) change in the same ratio as do the cylinder height and radius respectively*.
Hence

$$\cfrac{\cfrac{n_h}{n_{oh}}}{\cfrac{n_r}{n_{or}}} = \frac{h}{h_o} \cdot \frac{r_o}{r} = \lambda \cfrac{\dfrac{1}{\Pi}\left(\dfrac{V_o}{h_o}\right)^{1/2}}{\dfrac{1}{\Pi}\left[\dfrac{B(\lambda)\,V_o}{h}\right]^{1/2}} =$$

$$= \lambda^{3/2} \, B(\lambda)^{-1/2} \quad . \qquad \text{(see Eq. (5)).}$$

If we consider that

$$\frac{n_{or}}{n_{oh}} = \text{tg } \gamma_o \quad \text{and} \quad \frac{n_r}{n_h} = \text{tg } \gamma \, ,$$

we obtain the equation determining the angle which the director makes with the z axis after deformation

$$\text{tg } \gamma = \lambda^{-3/2} \, B(\lambda)^{1/2} \, \text{tg } \gamma_o \quad . \tag{6}$$

---

*It should be noted here that in the Kratky scheme the sample orientation was modelled by the orientation (at constant volume) of rod-like units whose symmetry axes rotated on drawing in an identical manner to the lines joining pairs of points in the uniaxially drawn sample, without any change of length of the rotating units. As true affine deformation assumption proposed by Kuhn and Grün for crosslinked elastomers[11], here quoted after[10], says that the effect of the deformation is to change the components of the vector length of each polymer chain in the same ratio as the corresponding dimensions of the bulk sample the Kratky scheme is termed pseudoaffine. In view of the above our treatment is affine deformation scheme for the vector field description of molecular orientation.

The probability of finding director lying at an angle between γ_0
and $\gamma_0 + d\gamma_0$ in indeformed volume element should be the same as
the probability of finding it between γ and $\gamma + d\gamma$ in this volume
element after deformation (the sample volume increased $B(\lambda)$ times
but the volume of the polymer remained unchanged).
Hence

$$a(\gamma_0) \sin \gamma_0 \, d\gamma_0 = a(\gamma) \sin \gamma \, d\gamma, \qquad (7)$$

where $a(\gamma_0) = 1$ random initial orientation distribution .
From Eq. (6) it follows that

$$\frac{d\gamma}{d\gamma_0} = \frac{1 + tg^2\gamma_0 \left[\frac{B(\lambda)}{\lambda^3}\right]^{1/2}}{1 + tg^2\gamma_0 \frac{B(\lambda)}{\lambda^3}} . \qquad (8)$$

Combination of Eq. (1) and Eqs. (6) ÷ (8) gives the orientation
distribution function for the sample increasing its volume $B(\lambda)$
times with the draw ratio:

$$f_f(\gamma,\lambda) = a_f(\gamma,\lambda) \sin \gamma = \qquad (9)$$

$$= \frac{\left[\frac{\lambda^3}{B(\lambda)}\right]^{1/4} \sin \gamma}{\{\left[\frac{\lambda^3}{B(\lambda)}\right]^{-1/2} \cos^2\gamma + \left[\frac{\lambda^3}{B(\lambda)}\right]^{1/2} \sin^2\gamma\}^{3/2}}$$

Halfwidth of a_f is equal to

$$\sin^2\gamma_f 1/2 = \frac{\sqrt[3]{4} - 1}{\frac{\lambda^3}{B(\lambda)} - 1} \qquad (10)$$

orientation parameter

$$S_f(\lambda) = \left[3R_f(\lambda) - 1\right] \qquad (11)$$

where

$$R_f(\lambda) = <\cos^2\gamma>_f = \frac{\frac{\lambda^3}{B(\lambda)}}{\frac{\lambda^3}{B(\lambda)} - 1} - \frac{\frac{\lambda^3}{B(\lambda)}}{\left[\frac{\lambda^3}{B(\lambda)} - 1\right]^{3/2}} \text{ arc cos } \left[\frac{\lambda^3}{B(\lambda)}\right]^{-1/2} \qquad (11')$$

The approach described above is a generalization of that proposed originally by Kratky[7]. Putting $B = const. = 1$ his, constant - volume, model for pure polymer is obtained. From Eqs. (10) and (11) it can be noticed that an increase of $B(\lambda)$ (e.g. due to the introduction of a filler into the polymer matrix or due to its higher concentration) causes an increase of $\gamma_{f1/2}$, and decrease of the orientation parameter $S_f(\lambda)$. We take this fact to result from the increase of the volume fraction of unoriented polymer matrix $k(\lambda)$ with the increase of the volume fraction of the filler in the sample. The values of the orientation distribution function for the strained elements of the polymer matrix $f(\gamma,\lambda')$ are obtained after putting in Eq. (9) $B(\lambda) = 1$ and the draw ratio

$$\lambda' = \frac{\lambda - k(\lambda)}{1 - k(\lambda)} \quad , \text{ (see Eq. (3))} . \qquad (3')$$

Now we demand that for each λ the value of the orientation parameter calculated from Eq. (11) should be equal to the value obtained by integration of Eq. (4) with the orientation distribution function for the oriented phase as described above. Basing on Eq. (4) the orientation parameter equals

$$S_{fs}(\lambda) = \tfrac{1}{2}\left[3R_{fs}(\lambda) - 1\right]$$

where

$$R_{fs} = {}_0\!\int^{\Pi/2} f_{fs}(\gamma,\lambda) \cos^2\gamma \; d\gamma =$$

$$= k(\lambda) {}_0\!\int^{\Pi/2} \sin\gamma \, \cos^2\gamma \; d\gamma + \left[1 - k(\lambda)\right] {}_0\!\int^{\Pi/2} f(\gamma,\lambda') \cos^2\gamma \; d\gamma =$$

$$= \tfrac{1}{3} k(\lambda) + \left[1 - k(\lambda)\right] \cdot R(\lambda')$$

and $R(\lambda')$ is obtained from Eq. (11') after putting $B(\lambda) = 1$ and λ', as in Eq. (3').

Demanding that

$$S_f(\lambda) = S_{fs}(\lambda)$$

leads to the condition

$$R_f(\lambda) = R_{fs}(\lambda)$$

or in a full from

$$\frac{1}{3} k(\lambda) + [1 - k(\lambda)]\{ \frac{[\frac{\lambda - k(\lambda)}{1 - k(\lambda)}]^3}{[\frac{\lambda - k(\lambda)}{1 - k(\lambda)}]^3 - 1} -$$

$$- \frac{[\frac{\lambda - k(\lambda)}{1 - k(\lambda)}]^3}{[\frac{\lambda - k(\lambda)}{1 - k(\lambda)} - 1]^{3/2}} \text{ arc cos } [\frac{\lambda - k(\lambda)}{1 - k(\lambda)}]^{-3/2} \} =$$

$$\tag{13}$$

$$= \frac{\frac{\lambda^3}{B(\lambda)}}{\frac{\lambda^3}{B(\lambda)} - 1} - \frac{\frac{\lambda^3}{B(\lambda)}}{[\frac{\lambda^3}{B(\lambda)} - 1]^{3/2}} \text{ arc cos } [\frac{\lambda^3}{B(\lambda)}]^{-1/2}$$

The results of numerical solution of Eq. (13) on the basis of the experimentally obtained $B(\lambda)$ dependence will be presented for various degrees of filling in one of the following sections.

Influence of unoriented volume fraction on tensile properties

In order to predict the stress-strain dependence for a polymer filled with a non-adhering filler on the basis of known stress-strain curve for pure polymer we have to find the deformed sample effective load bearing cross-section.
Let us notice that the load bearing capability of any body is determined by its weakest, i.e. smallest, cross-section. In the case of a volume element with initial dimensions (length, width, thickness) l_o, w_o d_o, filled with homogeneously distributed filler particles, its effective cross-section in undeformed state is reduced to the value

$$S_{oeff} = S_o - n(S_o) <S_{fo}> =$$

$$= (S_o - N^{2/3} \frac{S_o}{(l_o w_o d_o)^{2/3}} <S_{fo}>) =$$

$$= S_o (1 - f^{2/3} \frac{<S_{fo}>}{<V_g>^{2/3}}) = S_o \cdot r_f , \tag{14}$$

where: S_o – initial sample cross-section $S_o = d_o \cdot w_o$
       $n(S_o)$ – the maximum number of filler particles passing
                through the volume element ("sheet") with the length
                $\Delta l_o \ll l_o$ and area $d_o \cdot w_o$,

$<S_{fo}>$ - the maximum average area of the cross-section in d_o, w_o plane of one filler particle with the above mentioned volume element,

f - volume fraction of a filler $f = \frac{V_f}{V_o}$, V_f total volume of a filler in volume V_o element $V_o = l_o w_o d_o$,

$<V_g>$ - average volume of a filler particle.

After sample deformation to some value of the draw ratio λ, its effective cross-section $S_{eff}(\lambda)$ can be found from the constant matrix volume condition

$$\Delta 1 \; S_{oeff} = \Delta 1' \; S_{eff}(\lambda) \qquad (15)$$

where $\Delta 1$ - thickness of the considered "sheet" after deformation. Making use of Eq. (14) and Eq. (3) (in the strained areas of the sample the draw ratio is higher) we obtain

$$S_{eff}(\lambda) = \frac{S_o \cdot r_f}{\lambda'}$$

where - r_f is defined by Eq. (14), λ' is defined by Eq. (3), which can be rewritten in the form

$$S_{eff}(\lambda) = \frac{r_f \cdot S(\lambda)}{B(\lambda) \; \frac{\lambda'}{\lambda}} \qquad (16)$$

if we account for the sample volume increase during deformation i.e.

$$S_o \cdot l_o \cdot B(\lambda) = S(\lambda) \cdot 1$$

where $S(\lambda)$ - actual sample cross-section.

If the sample is loaded with the force $F(\lambda)$, the stress calculated per its effective cross-section equals

$$\sigma_{eff}(\lambda) = \frac{F(\lambda)}{S_{eff}(\lambda)} \quad , \qquad (17)$$

and it should be equal to the stress in pure polymer strained to the draw ratio λ', $\sigma_p(\lambda')$

$$\sigma_{eff}(\lambda) = \sigma_p(\lambda') \quad . \qquad (18)$$

From Eqs. (16), (17) and (18) we obtain the relationship determining the stress acting on the sample (load per sample cross-section)

$$\sigma_s(\lambda) = \frac{r_f}{B(\lambda) \cdot \lambda'/\lambda} \cdot \sigma_p(\lambda') \qquad (19)$$

This relationship will be tested experimentally in the next section.

EXPERIMENTAL

The experiments described below were used to verify the theoretical predictions made in the previous section.

Materials and samples

Isotactic polypropylene (i-PP) Malen J-400 and chalk of Polish production were used throughout this work. The average size of the chalk particles was calculated from their size distribution known from the previous works[2]. It was assumed that chalk particles are cylinder shaped with diameter h_1 = 1.2 μm and length h_2 = 8 μm. Oligomer of ethylene oxide (OEO) M_w = 300 was used as chalk modifier.

Compositions of polypropylene with unmodified and modified chalk were obtained by means of a screw extruder at 210° C. The granulated product was used to prepare compression moulded films 0.4 mm thick which were pressed at 190° C and then quenched in iced water. From films obtained in this way oarshaped samples with the dimensions as in Fig. 2 were punched. In all cases the samples were cut out form the same places in compression moulded films in order to provide approximately the same flow-induced orientation of chalk particles relative to the sample axis. For the purpose of tensile properties and volume changes measurements the samples were marked (see Fig. 2).

The following compositions were investigated:
- pure PP,
- 90 wt.% PP + 10 wt.% (90 wt.% chalk + 10 wt.% OEO 300) (referred to as 9:1),
- 70 wt.% PP + 30 wt.% (90 wt.% chalk + 10 wt.% OEO 300) (referred to as 7:3),

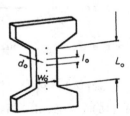

Fig. 2. The shape of the samples used in the tensile experiments. L_o = 10 mm, l_o = 2 mm, w_o = 5 mm, d_o = 0.4 mm.

- 60 wt.% PP + 40 wt.% (90 wt.% chalk + 10 wt.% OEO 300) (referred to as 6:4),
- 50 wt.% PP + 50 wt.% (90 wt.% chalk + 10 wt.% OEO 300) (referred to as 5:5),
- 40 wt.% PP + 60 wt.% (90 wt.% chalk + 10 wt.% OEO 300) (referred to as 4:6).

Methods

Studies of the tensile properties concerned true stress-true strain measurements taken by means of an INSTRON Tensile Testing Machine and video system. The course of the deformation process running at 20%/min. (2 mm/min.) rate was recorded by means of a video camera. Observations were performed on the volume elements of each sample (see Fig. 2). Changes of sample thickness were measured with a properly modified micrometer screw. Elasticity moduli E, elasticity limit ε_E and elongation at yield ε_y were determined on the basis of registered load - crosshead displacement curves. Relative change of volume of the observed elements of the samples, defined as the ratio of its actual to initial volume, was calculated for various draw ratios on the basis of sample dimensions measurements (see above). The degree of polymer matrix orientation was estimated from X-ray diffraction (Cu Kα line λ = 1.5405 Å, DRON - 2 diffractometer) measurements. As meridional reflections (001) are not available in the i-PP X-ray diffraction pattern[12], halfwidths of the intensity profiles of an equatorial reflection (110) were used as an empirical index of the degree of alignment. Azimuthal intensity profiles were obtained in transmission by step--wise sample rotation around the X-ray beam axis. As the presence of unoriented phase was supposed, the peak heights could not be measured from the minima in the azimuthal scan. Background measurements were carried out by determining diffraction intensities at the angles 2θ = 8° and 2θ = 31° as a function of sample rotation angle and interpolating the value for 2θ = 14° (110 reflection). After substraction of the background profile from the reflection profile, true peak shape was obtained. The location of the unoriented areas was determined from transmission electron microscopy studies (TESLA BS 500 microscope) of two-step carbon replicas taken from the surface of gold decorated and deformed samples. Both virgin sample surfaces and surfaces obtained by microtoming a thin section from the LN_2 cooled samples were studied.

RESULTS AND DISCUSSION

Fig. 3 presents the results of measurements of relative sample volume changes vs. draw ratio for various filler contents. As was mentioned in the previous section, the observations were carried out on sample volume elements with initial length l_0 = 2 mm.

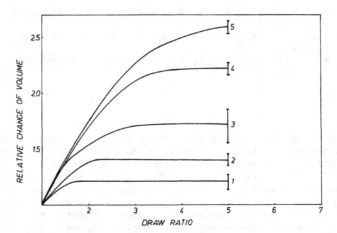

Fig. 3.   The results of measurements of relative sample volume
           changes vs draw ratio for various filler contents.
           (1 - 9:1,   2 - 7:3,   3 - 6:4,   4 - 5:5,   5 - 4:6).

The characteristic saturation of these curves is due to reorientation
of the filler particles and subsequent partial closing of voids[4].

     The results of numerical solution of Eq. (13) with such experi-
mentally obtained $B(\lambda)$ values in the form of unoriented volume
fraction of a polymer matrix vs. draw ratio curves for various filler
contents are shown in Fig. 4. As could be expected, unoriented volume
fraction of a polymer matrix for a given draw ratio λ increases with

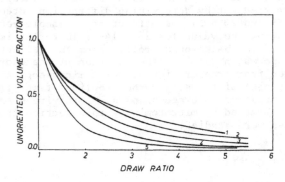

Fig. 4.   Calculated unoriented volume fraction of a polymer matrix
           vs draw ratio curves for various concentrations of a filler
           (1 - 4:6,   2 - 5:5,   3 - 6:4,   4 - 7:3,   5 - 9:1).

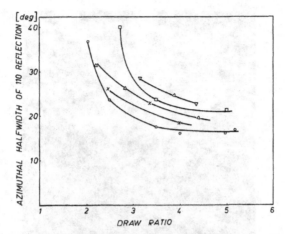

Fig. 5.   Azimuthal halfwidth of 110 reflection vs. draw ratio for
          various filler concentrations (O - 1:0,   × - 9:1,
          △ - 7:3,   □ - 5:5,   ▽ - 4:6).

increasing filler content and decreases with the draw ratio at a rate
which decreases with increasing filler content in the sample.   Fig. 5
shows the results of azimuthal halfwidth of 110 reflection vs. draw
ratio measurements.   It should be mentioned that the accuracy of these
measurements decreases for lower draw ratios due to the inaccuracy
of small λ determination, but the general trend of decrease of the
orientation parameter with the increase of filler content is evident.
This observation is in agreement with the results of calculations
presented in Fig. 4.

     Transmission electron micrographs of two-step PS-carbon replicas
taken from the surfaces of gold decorated samples deformed to the
various values of draw ratio and containing 40 wt.% of modified chalk
are presented in Fig. 6. Fig. 6a) shows a void around a filler particle in the early stage of deformation, the intermediate stage of
deformation is shown in Fig. 6b), while the situation just before
the sample fracture is presented in Fig. 6c). Presence of unoriented
fraction of polymer matrix at the pole sides of voids and its decrease
with the draw ratio, being in agreement with the theoretical predictions, is well documented by the above data.

     The results of measurements of some tensile properties in the
small strain range are presented in Fig. 7 in the form of curves
of E_F/E_P, $\varepsilon_{EF}/\varepsilon_{EP}$, $\varepsilon_{YF}/\varepsilon_{YP}$ vs filler volume fraction. E_F, ε_{EF}, ε_{YF}
and E_P, ε_{EP}, ε_{YP} are the moduli, elasticity limits, and elongations
and yield of filled and pure polymer samples respectively. The origin
of the maximum on the E_F/E_P curve can be explained if we consider

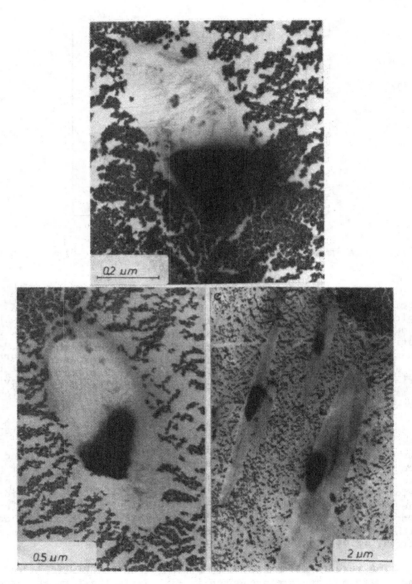

Fig. 6. Transmission electron micrographs of two-step PS-carbon
         replicas from the surfaces of gold decorated samples.
         a) initial, b) intermediate, c) final stages of defor-
         mation. Modified filler contents - 40 wt.%.

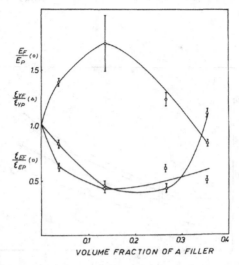

Fig. 7. E_F/E_P, $\varepsilon_{EF}/\varepsilon_{EP}$, $\varepsilon_{YF}/\varepsilon_{YP}$ vs volume fraction of a filler
curves (see text).

that an increase of the volume fraction of a filler in the sample
causes (see Eq. (19)):
-   an increase of the undeformed volume fraction and, by the same
    token, an increase of the draw ratio and stress in the strained
    areas and
-   a decrease of the effective sample cross-section.
For lower filler volume fractions the first effect is stronger
raising the sample modulus, while with the increase of filler content
the second one becomes dominating and the modulus drops. Let us
notice that the alternate explanation of the modulus increase for
lower - and its decrease for higher filler concentrations seems to
be invalid as it would require strong polymer filler adhesion in the
initial stages of deformation decreasing with the increase of filler
content. The shape of $\varepsilon_{EF}/\varepsilon_{EP}$ and $\varepsilon_{YF}/\varepsilon_{EP}$ curves can be explained
in the same terms. Rapid increase of the $\varepsilon_{YF}/\varepsilon_{YP}$ curve for the
highest filler volume fraction suggests some change in the mechanism
of yield in this case. This can be associated with the observation
that no neck propagation is then observed and deformation does not
localize in any place in the sample which deforms rather homogeneously
(as will be quantitatively shown later). The results of measurements
of tensile properties in the range of large strain are presented
together with the values of true stress calculated from Eq. (19) in
Fig. 8a) and 8b) in the form of true stress vs draw ratio curves.

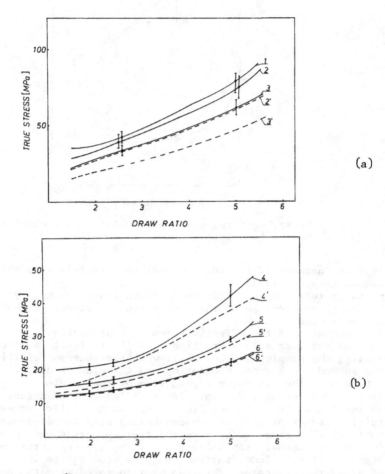

Fig. 8. Measured (solid lines) and calculated (broken lines) true
stress vs draw ratio curves for various concentrations
of a filler: a) 1 - 1:0,   2 - 9:1,   3 - 7:3;
b)   4 - 6:4,   5 - 5:5,   6 - 4:6.

The values of parameter r_f determing the effective sample cross--section

$$r_f = 1 - f^{2/3} \frac{\langle S_{fo} \rangle}{\langle V_g \rangle^{2/3}}$$

(see Eq. (14)) were calculated with the help of the following equation

$$\frac{\langle S_{fo} \rangle}{\langle V_g \rangle^{2/3}} = \begin{cases} \dfrac{(\Pi/4)^{1/3}(h_1/h_2)^{2/3}}{\langle \cos \theta \rangle} & \text{for } \langle \theta \rangle \, \varepsilon < 0; \text{ arc ctg } (h_1/h_2) \\[3mm] \dfrac{(\Pi/4)^{-2/3}(h_1/h_2)^{-1/3}}{\langle \sin \theta \rangle} & \text{for } \langle \theta \rangle \, \varepsilon < \text{ arc ctg } (h_1/h_2); \ \Pi/2 \rangle \end{cases}$$

derived for the case of cylindrical shape of filler particles, where h_1, h_2 – average cylindrical filler particle diameter and length respectively, θ – the angle between the long axis of a filler particle and the deformation direction, and $\langle \ \rangle$ – average over all possible values of θ. Random distribution of initial orientation of filler particles was assumed so the first equation with $\langle \cos\theta \rangle = \frac{1}{2}$ value was used. From Figs. 8a) and 8b) it is seen that the best agreement between experimental and theoretical results is achieved for higher filler concentrations (Fig. 8a)). One of the reasons for the discrepancy between the experimental results and theory is easily understood if it is kept in mind that all calculations were done with the assumption of homogeneous deformation of the considered volume element of a sample (A4). As was said earlier, volume elements with the length $l_o = 2$ mm were considered. The departure from the above mentioned assumption was tested in such a way that the heterogeneity parameter of sample deformation was defined as

$$\chi(\lambda) = \frac{\lambda}{(l_{os} + V_B \cdot t)/l_{os}}$$

where λ – draw ratio of the considered volume element with initial length l_o, l_{os} – initial length of the sample, V_B – crosshead speed t – time from the beginning of deformation and plotted for various filler contents vs. the draw ratio of the considered volume element λ (see Fig. 9). As can be seen, deformation becomes increasingly more heterogeneous as the filler volume fraction in the sample decreases. Thus the value of strain in the volume elements that can really be considered as deformed homogeneously is higher than the determined value which is responsible for the discussed experiment theory discrepancy in the range of small filler concentrations.

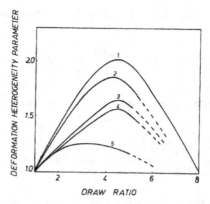

Fig. 9. The sample deformation heterogeneity parameter vs. draw ratio
         for various concentrations of a filler: 1 - 1:0,   2 - 9:1,
         3 - 7:3,   4 - 6:4,   5 - 4:6.

The experimental results presented above permit the conclusion that
the model presented in the previous sections satisfactorily accounts
for the tensile behavior of polypropylene filled with non-adhering
filler.   The best quantitative agreement between theory and experiment is obtained for the samples with higher volume fractions
of the filler which deform in a more homogeneous way.

CONCLUSIONS

      Qualitative consideration of the deformation process of polymer
filled with non-adhering particulate filler leads to the conclusion
that, depending on the amount of filler and on the draw ratio some
unoriented volume fraction $k(\lambda)$ should be present in the strained
filled sample. Basing on the assumption that unoriented areas ought
to be located at the pole sides of voids around the filler particles
and using the experimentally found parameter, a semiempirical model
was proposed, the first part of which made it possible to calculate
the values of $k(\lambda)$ for various degrees of filling. The second part
of the model gave the stress vs. draw ratio curves for the considered
polymer-filler system, making use of known true stress vs draw ratio
curves for pure polymer matrix. In all calculations homogeneous
deformation of the described volume element of the sample was assumed
(the value of λ does not depend on position in it). All these predictions were then tested experimentally. Tensile properties at small
and large strain range, TEM and X-ray difraction experiments confirmed
theoretical predictions. Some discrepancy between the experiment and
theory in large strain range tensile properties is explained in terms
of departure from deformation homogeneity of the considered volume
element in low filler volume fraction samples. This argument is

supported by deformation heterogeneity measurements which show it
to increase with a decrease of filler volume fraction.

REFERENCES

1. A. Gałęski, R. Kaliński, in "Polymer Blends. Processing,
    Morphology and Properties" eds. E. Martuscelli, R. Palumbo
    and M. Kryszewski, Plenum Press, New York, London (1980),
    p. 431.
2. R. Kaliński, A. Gałęski, M. Kryszewski, J.Appl.Polym.Sci.,
    26:3982 (1981).
3. B. M. Badran, A. Gałęski, M. Kryszewski, J.Appl.Polym.Sci.,
    27:3669 (1982).
4. T. Kowalewski, R. Kaliński, A. Gałęski, M. Kryszewski,
    Colloid and Polym.Sci., 260:652 (1982).
5. J. A. Manson and L.H. Sperling, "Polymer Blends and Composites",
    Plenum Press, New York (1976).
6. J. Schulz, "Polymer Materials Science", Prentice Hall,
    Englewood Clifts (1974).
7. O. Kratky, Kolloid-Z., 64:213 (1933).
8. R. Zbinden, "Infrared Spectroscopy of High Polymers",
    Academic Press, New York and London (1964).
9. I. M. Ward, "Mechanical Properties of Solid Polymers",
    Wiley Interscience, London (1971).
10. D. W. Hadley and I.M. Ward, in "Structure and Properties of
    Oriented Polymers" ed. I.M. Ward, Appl.Sci. Publishers,
    London (1975), pp. 264-289.
11. W. Kuhn, F. Grün, Kolloid-Z., 101:248 (1942).
12. R. J. Samuels, "Structured Polymer Properties", Wiley Inter-
    science, London (1971).
13. B. Wunderlich, "Macromolecular Physics", Vol. 1, Academic Press,
    New York (1973).

COMPOSITE MEMBRANES CONTAINING LIQUID CRYSTALS -

- SOME THERMAL AND MECHANICAL PROPERTIES

M. Kryszewski, Z. Bartczak

Centre of Molecular and Macromolecular Studies
Polish Academy of Sciences
90-362 Łódź, Poland

INTRODUCTION

Pressure or concentration difference serves as the driving force for isothermal diffusion through polymer films. Molecules at higher concentration sorb into the solid polymer and can move through the matrix of polymer chains with subsequent desorption from the film surface of different concentration. If the polymer membrane separating two solutions is permeable, the diffusion of molecules from the high concentration side to the low concentration side is called osmosis or dialysis.

If the molecules are made to diffuse from the low concentration side of polymer membrane to the high concentration side, this process is called reverse osmosis. This can be achieved by applying to the low concentration side a pressure high enough to overcome the opposing force (osmotic pressure) that would normally drive the molecules from a high concentration into a low one. Reverse osmosis is an important application of diffusion through polymers. Currently many efforts are made to use diffusion through polymer films for separation of different substances: ions, gases, vapours and liquids.

There are two ways of looking at how the molecules diffuse through a solid polymeric material. Barrer[1,2] is taking into consideration the fluctuation of the thermal energy in the polymer. When the local energy is sufficiently high, the diffusing molecules can move or jump between polymer chains by a cooperative motion with them. In that theory, it is assumed that polymer chain segments are involved in making the opening and the larger the diffusing

243

molecule, the greater the number of polymer segments involved in
this diffusion step.

The other approach derives from the free volume theory of
diffusion[3]. In this point of view, it is assumed that the fluctuations of local density in the polymer result in free volume or
hole formation. When the hole is of sufficiently large size, and
if it is formed near the penetrant molecule, the molecule can move
or jump into it. The diffusion is assumed to be proportional to
the probability of forming such holes of right size. The effect
of the diffusing molecules on the free volume formation can be taken
into consideration, too.

Both theories of diffusion imply that the number of polymer
segments involved in the diffusion step increases with the size
of the diffusing molecule.

Calculations based on these theories provide the basis for
correlating the diffusion data and for determining the physical
significance of the diffusion coefficient.

Many monographs (see e.g.[4]) and reviews have been devoted to
the general theories of diffusion through polymer films as well as
to the relation of the diffusion coefficients D, permeability P,
and solubility S to polymer structure and morphology of the membrane.
Without going into detail it seems important for future discussion
to give some general remarks on these factors.

It has been found[5-7] that the solubility constant S (for rather
small molecules) is directly proportional to the amorphous phase
content. This effect is correlated with greater chain mobility in
amorphous regions. It has also been shown that as the size of the
penetrant molecules increases less of the total amorphous phase may
be accessible to it. The diffusion coefficient D for a given molecule diffusing through a polymer depends on crosslinking, chain
stiffness, crystallite size and distribution. Small crystallites
homogeneously distributed in the polymer create a more tortous
channel for the diffusing molecules in which they can move. Greater
path length results in lower values of the diffusion coefficients.
The presence of the crystalline phase restricts also the motion of
polymer segments causing the diffusion to decrease. The effects of
crystalline phase are usually characterized by geometric impedance
factor and chain immobilization factor. The diffusion through polymer
films is affected by the glassy or rubbery state of the polymer
matrix and by stretching of polymer specimen[8].

In the case of partially crystalline polymers many factors make
the correlation of diffusion data with polymer structure very difficult, and the general theories of diffusion can be applied only to
amorphous polymers in rubbery state (ordinary Fick's law), and

penetrant transport in polymeric glasses appears to be anomalous.
Time dependent boundary condition, relaxation, and crazing are the
phenomena which greatly influence the diffusion of low molecular
weight substances in polymer glasses. It seems important to note
that the permeability values taken from various sources should be
taken into consideration with care because sample preparation procedures may have been different and the morphology may vary to
a high extent.

There are numerous citations in the literature describing the
attempts to alter the physical or physicochemical structure of
polymer membranes to improve their properties as permselective
barriers. Most of them consist in elaborate membrane fabrication
schemes employing complex casting procedures or controlled swelling
in suitable liquids for controlled time periods at various temperatures. Preparation of polymer blends appears to be a very interesting
approach to changing and improving the membrane properties. The
permeation of gases and liquids through multicomponent polymer films
has been investigated by many authors taking into consideration the
chemical structure of components, state of aggregation, and morphology. Many interesting examples of permeation properties of polymer
blends are discussed in the monograph of Paul and Newman[9]. In
order to elucidate the membrane behaviour of polymer blends it is
important to know the details of compatibility or domain structure
formation (see e.g. the studies on compatibility of frequently
investigated system PPO-aPS using different techniques[10]) and the
influence of the membrane preparation procedure (see e.g. investigations on the influence of PPO-aPS film annealing and swelling on
transport properties[11]). The influence of thermal molecular mobility
on permeation in multicomponent systems is of as great importance as
in one component systems. It has been shown that the permeability
of some liquids and gases abruptly increases in the temperature
range of the onset of primary relaxation process in the component
which shows the main relaxation phenomena at lower temperature.

From that point of view it seems natural that special consideration should be given to the composite membranes in which one component consists of liquid crystal. The transitions from crystalline
to mesophase state are connected with mobility increase thus
influencing the transport properties of these composite systems.
However, it was only recently that a discontinuous jump of permeability to liquids, gases and vapours in the vicinity of transition
temperature of the liquid crystal phase has been discovered[12].
Membranes of this type form a new class of composites which deserve
special consideration due to their particular properties.

The aim of this work is to review the transport properties of
composite polymeric membranes containing liquid crystalls, and to
investigate some of their thermal and mechanical properties.

PERMEABILITIES IN POLYMER/LIQUID CRYSTAL COMPOSITE MEMBRANES

Recently very interesting results have been reported on the behaviour of polymer/liquid crystal membrane consisting of bisphenol A polycarbonate (PC) and of N-ethoxybenzylidene 4'-n-butyl aniline (EBBA)[13]. This substance shows a crystal-nematic transition at 304° K and nematic-to-isotropic phase transition at 355° K.

Two-component PC/EBBA membranes containing 15, 30, 45, and 60 wt.% of EBBA were prepared by casting from 1,2-dichloroethane solutions and dried in vaccum at about 355° K for several days. Homogeneous dispersion of EBBA was found in the range of lower concentrations of liquid crystal while for the 45 and 60 wt.% blends the system is characterized by the existence of crystalline domains[14] as wide angly X-ray sattering and DSC measurements. The DSC investigations have shown that the two transitions for EBBA are observed also for PC/EBBA blends containing 45 and 60 wt.%, while the blend with EBBA of 15 wt.% does not exhibit such transitions. This behaviour confirms the previously mentioned results of structural studies. Studies on the density-EBBA fraction relationship at low concentration of the latter show that the distance between the PC chains increases continuously which is an argument for PC and EBBA miscibility. The crystalline domains, on the other hand, exhibit transitions from crystalline to mesophase state playing an important role in the transport of gases and vapours.

Permeability studies were carried out by volumetric methods and the sorption of hydrocarbons was measured by thermogravimetric method. Membranes with higher concentration of EBBA seem to be more interesting from the point of view of permeability because they exhibit particular behaviour related to the changes occuring in the mesophase of the liquid-crystalline component.

The first studies on the permeation of these composite membranes concerned the diffusion of water[14].

The diffusive permeability coefficient P was determined from the hydraulic permeability coefficient K measured in a temperature range of 273-333° K according to the method described by Yasuda and Peterlin[15]. The ln P - 1/T plot shows a distinct jum at about 304° K for membrans with 45 and 60 wt.% of EBBA. It corresponds to the crystalline-nematic phase transition of EBBA. This important increase of P values is probably due to the transition in EBBA which may induce the activation of molecular motion of PC or to the creation of larger vacancies fraction around the domain boundary of EBBA. In that way the water molecules may diffuse faster. Pure PC films do not exhibit that type of behaviour, and the temperature dependence of P does not show, in that temperature range, any significant changes.

Interesting support for the above suggestion on the mechanism of the increase of P values has been obtained in the work of Maemura et al.[14]. The dynamic loss tangent (tan δ) measurements carried out in wet conditions (water bath) show a sharp peak at about 304° K for the 60% EBBA blend. This peak can be attributed to the onset of molecular motion induced by the transitions in EBBA domains. This effect was not observed in pure PC; the blend containing only 15 wt.% of EBBA exhibited only a gradual increase of tan δ. Water permeation through composite membranes with liquid crystals was correlated to the thermal molecular motion in polymeric matrix in analogy with one-component systems. Similar effects have been observed in polymeric membranes made of polymers which are frequently used as models for biopolymers e.g. poly(γ-methyl-L-glutamate)[16]. These results provided encouragement for the studies of permeation of gases and organic vapours in composite membranes containing liquid crystals.

Studies on the sorption of some hydrocarbons[17] have shown that above the transition temperature of EBBA (331° K) the isotherms obey Henry's law and the solubility coefficients S can be calculated. The sorption and desorption curves are similar in shape which indicates that these systems follow Fickian sorption. This fact indicates that steady state surface equilibrium is reached and that the diffusion coefficient for hydrocarbons is a function of concentration only. It follows that the membranes containing 60 wt.% of EBBA are homogeneous from the view point of gas permeation at the temperature above transition in EBBA. The permeability coefficients P show a distinct jump in the vicinity of transition temperature from crystal to nematic phase. This phenomenon was observed for hydrocarbon gases, noble gases like He, and for inert gases like N_2.

This effect is clearly seen from the plots of the function ln P = f(1/T). The differences in the slope of this function above and below the crystal-nematic liquid phase transition suggest that different mechanisms are involved in the diffusion of small molecules through these membranes. The slope of ln P - 1/T plot for pure PC membrane is constant because there are no transitions in the temperature range under investigation.

In the case of pure PC membrane, the values of permeability coefficients are larger for CH_4 than for C_3H_8. For the composite membrane containing 45 wt.% the reverse is true, i.e. for molecules with greater number of carbon atoms higher permeability coefficients were found above the transition temperature of the liquid crystalline component. These effects appear more clearly in membranes containing 60 wt.% of EBBA. From the ln P - 1/T plots it can be seen that below the transition of the liquid crystalline component the values of permeability coefficients are lower for bigger molecules than for smaller ones. These studies lead to the conclusion

that below the transition temperature the permeation process is
diffusion controlled and that above the transition temperature in
EBBA it depends on the solubility factor. This type of behaviour
of composite membranes seems to be of special interest and needs
to be discussed.

The membranes with 60 wt.% of EBBA can be treated, from the
view-point of permeability, as homogeneous media. In that case the
coefficients P, D and S are interrelated according to the simple
relation $P = D S$. The values of diffusion coefficients can be
easily obtained from the time lag and membrane thickness measurements.
The lg D vs. 1/T plots show a very distinct change of D in the
vicinity of the crystalline-to-nematic phase transition temperature.
The S values exhibit a similar increase in the transition region.
These observations show that the drastic change of P is related
to the increase both of D and S values. The onset of the mobility
increase must be related to the induced molecular motions in the
nematic phase of liquid crystal component and to the creation of
some freedom for PC segments translations and/or rotations leading
to specific interactions between permeates and the PC/EBBA composite
membrane. This remark, which seems to be valid for hydrocarbons,
is true for He and N_2 which are inert gases. The solubility parameter S is constant in that case but the values of the diffusion
coefficient markedly increase in the transition region.

It seems appropriate to finish this short survey of the
behaviour of such composite membrans with a remark that the
permeation process of gases, vapours and liquids may be controlled
by concentration of liquid crystal by phase structure and by temperature to a different degree in comparison with other polymeric
multicomponent systems. There is no doubt that these composite
membranes will attract increasingly more attention in the near
future.

SOME ASPECTS OF THERMAL PROPERTIES AND MECHANICAL BEHAVIOUR
OF COMPOSITE MEMBRANES CONTAINING LIQUID CRYSTALS

In practical applications composite membranes containing liquid
crystals are subjected to stresses which can be considerable. They
can also be applied at different temperatures. These factors may
alter their properties to some extent. In order to obtain some
information on the thermal and mechanical properties of these
systems we have investigated:
a) the Tg of these systems with different EBBA concentration in
    order to elucidate the interactions between liquid crystals
    molecules and polymer matrix molecules,
b) the mechanical properties in order to obtain some information
    on the behaviour of these systems under stress.
The composite films containing different weight percentages of EBBA

and bisphenol A polycarbonate were cast from methylene chloride
solutions. The cast films were dried in vacuum at 340° K for four
days. The conditions under which these films were made were
analogous to those used by Japanese scientists whose permeability
studies on these systems were discussed before.

On the basis of polarizing microscopy and transmission electron
microscopy (TEM) carried out at room temperature is can be concluded
that EBBA exists in homogeneous dispersion when its concentrations
are below 30 wt.%. Density measurements show an unexpected, high
miscibility of EBBA with PC at lower concentrations. Studies of
the morphology of films containing 30, 40 and 50 wt.% of EBBA show
the appearance of well shaped EBBA crystals. Optical micrographs
(Fig. 1 and 2) display many small EBBA crystals rather homogeneously
dispersed in the PC matrix. One can also see bigger crystals whose
dimensions may reach as much as 200 μm. This heterogeneous structure
of EBBA blends with PC was confirmed by the studies using TEM
(Fig. 3). Well-separated EBBA crystals exist in a continuous PC
matrix.

THERMAL PROPERTIES

Studies of Tg values give valuable information on the molecular
mobility of polymer chain segments and on the influence of additives
on the main and subsidiary relaxation processes in PC[18-22]. Special
consideration was given to the suppression and shift of the secondary
relaxation process occuring at about 170° K as well as to the influence of various diluents e.g. chlorinated bisphenyls and chlorinated
terphenyls[23-27] as well as compounds which are able to form charge
transfer complexes (see e.g.[28-30]). The Tg values studied in this
work concern mostly higher temperature range, where phase transition
appears, and the subsidiary relaxations have been investigated only
to a small extent because the particular permeation behaviour of
the composite membranes discussed before was found above room temperature, at which only the broad shoulder of the secondary
relaxations appears.

The glass transition temperature was determined using the
dilatometric method. Volume dependence on temperature was measured
on a rising temperature scale with a heating rate of 2° K/min. The
films were carefully degassed at room temperature before filling
the dilatometer with mercury.

The dependence of Tg on the concentration of EBBA shows the
expected decrease of Tg values at lower EBBA concentrations where
the liquid crystalline additive exists as a homogeneous solution in
PC. EBBA shows a plasticizing effect which is less pronounced than
that observed for typical plasticizers like dibutyl phthalate and
chlorinated phenyls. The blends of these compounds were prepared

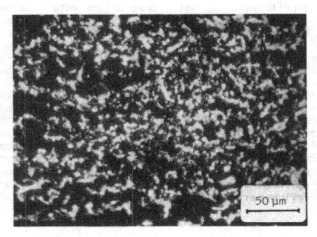

Fig. 1.   Optical micrograph of the system PC:EBBA = 50:50
          (with crossed polarizers).

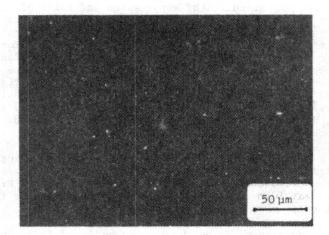

Fig. 2.   Optical micrograph of the system PC:EBBA = 70:30
          (with crossed polarizers).

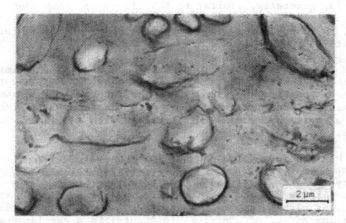

Fig. 3.   TEM micrograph of the free surface of the system
           PC:EBBA = 50:50.

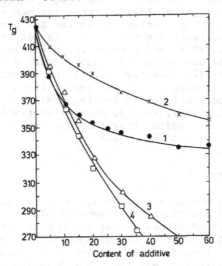

Fig. 4.   Glass transition temperature dependence on the EBBA content
           for PC-EBBA system (curve 1). For comparison the plots Tg
           vs. contents of chlorinated terphenyl, chlorinated diphenyl
           and dibutyl phthalate in PC also are presented (curves 2, 3
           and 4 respectively).

and investigated under the same conditions as the membranes
containing liquid crystalline additives. The decrease of Tg caused
by EBBA is, generally, similar to that found for compounds classified as antiplasticizers e.g. chlorinated biphenyls (Fig. 4).
However, the change of Tg values is more pronounced than for
chlorinated terphenyls.

For EBBA concentrations higher than 30 wt.%, a leveling of
Tg values can be easily detected. This behaviour may be correlated
with phase separation. In this case the interactions between EBBA
molecules and PC segments are limited to the interphase, thus leading
only to a very slow decrease of Tg values with additive fraction
increase.

It has been found that in the low temperature range the dynamic
mechanical loss is shifted to higher temperatures and reduced in
intensity when the additives act as antiplasticizers. This effect
also was observed in the case of PC for additives showing specific
interactions with polymer chain segments. The antiplasticizing and
plasticizing effect of additives (diluents) is a complex phenomenon.
Its elucidation in relation to molecular interactions is still far
from being complete, and will not be analysed here in detail. Our
studies of the mechanical loss in composite membrane films containing
up to 20 wt.% of EBBA, carried out in the low temperature range, have
shown an antiplasticizing activity of this additive. This effect
can be due to rather strong dipole-dipole interactions between
highly polar EBBA molecules and polar groups of PC segment, when
the liquid crystalline additive exists in molecular dispersion.
This activity disapears when phase separation occurs, i.e. for
higher EBBA concentrations. In the temperature range where secondary
relaxation in PC occurs, EBBA at high concentration exists in the
blend as a dispersed solid thus the interactions with PC chains
are limited only to the interphase.

MECHANICAL PROPERTIES

In order to characterize the basic mechanical properties of
the investigated materials, measurements of Young's modulus E,
stress at break, and elongation at break were investigated in the
function of EBBA content in the blend. Studies of the mechanical
properties were carried out on samples from 50 to 100 m thick
using an Instron Testing Machine. The relative elongation measurements at the rate of 10%/min., were made at ambient temperature
(20° C) i.e. below the solid-nematic transition of the liquid crystalline additive. The results obtained are represented in Figs. 5-7.
It follows from these figures that for the EBBA content below
20 wt.% an increase of modulus, E, and of stress at break appears
while the elongation at break ε_z exhibits a minimum in this range
of EBBA concentrations. These effects prove the antiplasticizing

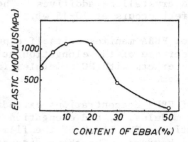

Fig. 5. Dependence of Young's modulus E on the concentration
of EBBA for the system PC:EBBA.

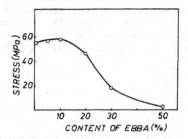

Fig. 6. Dependence of stress at break σ_z on the concentration
of EBBA for the system PC:EBBA.

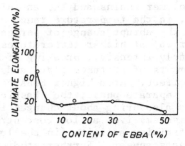

Fig. 7. Dependence of elongation at break ε_z on the concentration
of EBBA for the system PC:EBBA.

action of the liquid crystalline additives. The maximum of this antiplasticizing effect occurs at 10-15 wt.% of EBBA.

The influence of EBBA manifests itself especially in the appearance of the decrease of the elongation at break which shows that the additive interacts with PC molecules increasing the stiffness of the system.

In the range of EBBA concentrations exceeding 30 wt.%, a dramatic decrease modulus, E, and elongation at break appears due to phase separation. The break of the films under study occurs along the interphase borders. The above discussed morphology and the change of mechanical properties at EBBA concentrations exceeding 30 wt.% is connected with the jump of the permeation coefficients for gases and vapours[14]. At 20 wt.% of EBBA the antiplasticizing effect disappears, probably leading to the change of activation energy for permeability of vapours and gases as described by Maemura et al.[16]. It can be concluded that in the range of these concentrations the molecular dispersion is not preserved and some molecular aggregates of liquid crystalline additive appear. They are too small to be detected using microscopic techniques. With further increase of EBBA content, phase separation appears and the systems become heterogeneous.

CONCLUSIONS

The results of the study reported here suggest that the observed antiplasticizing process is connected with the bonding of PC chain elements by the polar molecules of the liquid crystalline additive. Thus at a sufficiently low concentration of EBBA at which molecular dispersion exists and at low temperatures this additive acts as an antiplasticizer. If, however, the content of the liquid crystalline phase increases and phase separation occurs, the inclusions of EBBA will only separate polymer chains and the antiplasticizing effect disappears and leads, in the temperature range of crystalline-nematic phase transition, to the abrupt change of permeation coefficient values. In a similar way at higher temperatures, when thermal motions are sufficiently intensive, breaking of the weak bonds between EBBA and segments of PC takes place and the additive will have a plasticizing effect. At a higher EBBA concentration, at which it exists as a separate phase, the plasticizing effect levels off. These phenomena also bring about changes in the mechanical properties of these systems i.e. at low concentrations EBBA molecules exhibit and antiplasticizing activity (in the low temperature range) and plasticizing effects appear. Further studies of these effects and their correlation with particular permeation using other liquid crystalline substances dispersed in conventional polymers are in progress.

REFERENCES

1. R. M. Barrer, "Diffusion in and through Solids" Cambridge University Press, Cambridge, (1941).
2. R. M. Barrer, J.Phys.Chem., 61:178 (1957).
3. C. A. Kumins and T.K. Kwei "Diffusion in Polymers", Academic Press, New York (1968).
4. J. Crank and G.S. Park "Diffusion in Polymers", Academic Press, New York, (1968).
5. A. W. Myers, C.E. Rogers, V. Stannett and M. Szwarc, Modern Plastics, 34:157 (1957).
6. A. S. Michaels and R.B. Parker, J.Polym.Sci., 41:53 (1959).
7. A. S. Michaels and J.J. Bixler, J.Polym.Sci., 50:393 (1961).
8. H. J. Bixler and A.S. Michaels "Effect of Uniaxial Orientation on the Liquid Permeability and Peremselectivities of Poly-olefins" 53 rd National Meeting American Institute of Chem. Eng. Pittsburgh, (1964), Preprint 32d.
9. H. B. Hopfenberg and D.R. Paul in "Polymer Blends", Eds. D.R. Paul and S. Newman, Academic Press, New York (1978) vol. 1, p. 445.
10. J. Stoelting, F.E. Karasz and W.J. Mac Knight, Polym.Eng. and Sci., 10:3 (1970).
11. C. Carfagna, A. Apicella, E. Drioli, H.B. Hopfenberg, E. Martuscelli and L. Nicolais in "Polymer Blends, Processing Morphology and Properties". Eds. E. Martuscelli, R. Palumbo and M. Kryszewski, Plenum Press, New York and London, (1980), p. 383.
12. T. Kajiyma, Y. Nagata, E. Maemura and M. Takayanagi, Chemistry Letters, (Japan),:679 (1979).
13. Y. Nagata, T. Kajiyma and M. Takayanagi, Rept.Progr.Polym.Phys. (Japan), 23:311 (1980).
14. E. Maemura, Y. Nagata, T. Kajiyma and M. Takayanagi, Rept.Progr. Polym.Phys. (Japan), 21:221 (1978).
15. H. Yasuda and A. Peterlin, J.Polym.Sci., Phys.Ed., 9:1117 (1971).
16. E. Maemura, Y. Nagata, T. Kajiyama and M. Takayanagi, Rep.Progr. Polym.Phys. (Japan), 20:663 (1977).
17. Y. Nagata, T. Kajiyma and M. Takayanagi, Rep.Progr.Polym.Phys., (Japan), :257 (1979).
18. L. M. Robeson and J.A. Faucher, J.Polym.Sci., Part B, 7:35 (1969).
19. L. M. Robeson, Polym.Eng.Sci., 9:227 (1969).
20. S. E. B. Petrie, R.S. Moore and J.R. Flick, J.Appl.Phys., 43:431 43:4318 (1972).
21. M. G. Wyzgoski and G.S.Y. Yeh, Polymer J., 4:29 (1973).
22. G. Pezzin, A. Ajroldi and G. Garbuglio, J.Appl.Polym.Sci., 11:2553 (1967).
23. S. E. Petrie, J.Polym.Sci., 10:1255 (1972).
24. W. J. Jackson Jr. and J.R. Coldwell, J.Appl.Polym.Sci., 11:211 (1967).

25. W. J. Jackson Jr. and J.R. Coldwell, J.Appl.Polym.Sci., 11:227
    (1967).
26. R. M. Mininni, R.S. Petrie and S.E.B. Petrie, Bull.Am.Phys.Soc.,
    17:272 (1972).
27. R. M. Mininni, R.S. Moore, J.R. Flick and S.E.B. Petrie,
    J.Macromol.Sci. Phys., B8:343 (1973).
28. L. Makaruk, H. Polańska, Polimery (Polish), 19:113 (1974).
29. L. Makaruk, H. Polańska and E. Stratos, J.Appl.Polym.Sci.,
    20:63 (1976).
30. M. Kryszewski and J. Ulański, Ann.Sci.Chim.Polonorum, 50:1441
    (1976).

STRUCTURE AND PROPERTIES OF TWO-COMPONENT MEMBRANES MADE
OF IONIC/NONIONIC POLYMERS WITH GRADIENT CONTENT OF AN
IONIC COMPONENT

A. Narębska, R. Wódzki and Z. Bukowski

Institute of Chemistry
N. Copernicus University
87-100 Toruń, Poland

Two-component membranes were synthesized by volume
grafting of acrylic acid on polypropylene in the form of
thin films. Distribution of acrylic acid across the membranes was determined by microphotometration of the
coloured thin cross-section of the membranes cut perpen-
diculary to the surface. It was found that grafted polyacid produces a concentration gradient inside the membranes.
Applying the multilayer electrochemical model and conductivity data the volumes of both components, i.e., PP and
swollen gel of g-PAA in the gradient membranes, were
calculated.

INTRODUCTION

The polymer films or membranes used as ionic conductors are
often made of composite polymeric materials made up of inert and
ionic polymers. In such a material the inert polymer ensures the
necessary elasticity and mechanical stability of the membranes,
whereas the ionic component is responsible for the electrochemical
properties.

As the components are incompatible they form some kind of dispersion. Mutual distribution of the polymers in the membrane depends
very much on the method of preparation and may vary from random,
when the membrane is made of e.g. copolymer grafted in solution,
up to gradient when the ionic component penetrates the film of the
inert polymer by means of diffusion and is then polymerized.
However, for the preparation of the gradient material it is

257

necessary for the rate of local grafting or interpolymerization to exceed that of diffusion.

For reasons of mechanical stability and separation ability the gradient material could be comparable to or better than that in which distribution of the ionic component is statistical.

In our studies on the preparation and structure of two-component membranes we followed one of the methods described by Pegoraro[1] choosing polypropylene as the matrix and polyacrylic acid as the ionic polymer.

MEMBRANES

The membranes were prepared by volume grafting of acrylic acid onto polypropylene activated by "soft" oxidation of the polymer in the presence of tert-butylhydroperoxide as the initiator and sodium dodecylsulphate as the surface active agent. During the preparation, samples of preactivated polypropylene film were immersed in aqueous solution of acrylic acid of 1.84 mol/dm^3 concentration for periods of time ranging from 4 up to 14 hrs. The monomer penetrating the film by sorption was polymerized within the film at 358 K.

As was found by microphotometration, the concentration gradient of the grafted PAA inside the membrane and its profile depend very much on the amount of the monomer absorbed. It was also evident that, because of the method used for preparation, polymerization of the acid was initiated mainly by PO· radicals formed on polypropylene chains by thermal and redox decomposition of POOH hydroperoxides or POOP peroxides. The homopolymer of polyacrylic acid which was also formed inside the film was later removed by long time elution with methanol.

The electrochemical and mechanical properties of the membranes containing different amounts of grafted PAA can be seen in Table 1. By test examinations of the membranes it was found that separation abilities in alkaline silver/zinc batteries are shown only by those in which the amount of g-PAA ranges from about 20 to 30 volume percent. The corresponding experimental data are presented in Table 1.

CONCENTRATION PROFILES OF PAA IN THE MEMBRANES

To investigate further the composition and structure of the membranes more experiments were performed.

The method which supplies straightforward information about the distribution of the ionic component in membranes is microphotometration of the thin cross-sections of the membranes coloured with

Table 1. Electrochemical and mechanical properties of PP-g-PAA membranes

Percentage of grafting wt. % in dry state	Conductivity in 0.1 KOH solution ohm^{-1}m^{-1}	Tensile strength at break (wet membrane) MPa	Elongation al break (wet membrane) %	Number of repeated charge-discharge cycles of AgO/Zn battery
11	-	28	390	-
20	0.0106	27	384	-
28	0.91	22	372	35
38	1.45	20	305	30
41	1.62	19	287	25
50	2.10	16	151	17
66	3.20	12	47	6

an appropriate dye. This method was used in our studies. Methylene blue was chosen for dying the membranes. This is a cationic dye which is absorbed in the membrane by ion exchange with the counterions of carboxylic groups and for this reason appears only in the domains containing the ionic component. The equipment[2] used for photometration consisted of a microscope, a photomultiplier connected to a synchronous drive mounted on the microscope stage and a recorder which registered automatically the intensity of the transmitted light along the scanned cross-section. The corresponding curves are shown in Fig. 1.

The concentration profiles show the distance to which polyacrylic acid penetrated the different samples. At the top of each curve there are figures which correspond to the volume fraction of the grafted polyacid in the dry membranes determined independently by a weighing technique. In calculations of the volume fractions of grafted polyacid, the density of PP was assumed to be equal to 0.898 g/cm^3 and that of PAA 1.441 g/cm^3.

It can be seen that when the volume fraction of PAA is below 13% it does not form a continuous phase throughout the membrane. Also the resistance of this membrane is much too high to treat it as the conducting material. When the volume fraction of PAA exceeds 13% the profiles shift forward and finally – at about 20% – the polyacid reaches the center. With increasing amount of PAA, the gradient slowly disappears and above 60% it is more or less uniformly.

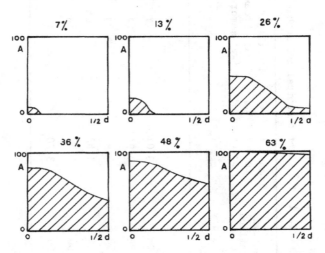

Fig. 1.   Concentration profiles of polyacrylic acid in two-component PP-g-PAA membranes determined by photometration. Light absorption expressed in relative units $100A_x/_{63\%, d=0}$.

distributed in the membrane. However, with the amount of PAA as
high as 60% the membrane swells strongly and its mechanical properties become really poor. Thus only membranes containing around 20
but less than 30 volume percent of PAA have practical importance.

PHASE COMPOSITION OF PP-g-PAA MEMBRANES ACCORDING TO MULTILAYER ELECTROCHEMICAL MODELS

When used in different kinds of electrochemical equipment the
membranes are in contact with aqueous solutions of the low molecular
weight electrolytes in which they swell. Moreover, a certain amount
of the ambient solution penetrates the voids or pores in the membrane.
So the swollen membrane is a multiphase system composed of an ion
containing component appearing in a gel state, an inert partly crystalline polymer, and the electrolyte filling any voids or nonselective domains, all of them in varying amounts. For such a system it
is possible to calculate the approximate phase composition based on
the conductivity and the multilayer electrochemical model. We presented such a model at the First Italian-Polish Seminar on Multicomponent Polymeric Systems[3] in 1979.

In the geometric representation of the composition of the
membrane, the volume fraction of each component in each layer is
described quantitatively by corresponding geometric parameters. In
the electrochemical part of the model, each layer is treated as a
set of two resistors and all sets, whose number equals the number
of layers, are arranged in series forming an equivalent electrical
circuit. Summation of the resistances of the layers expressed with
appropriate equations leads to the final formula on specific conductivity of the membrane. Calculations based on the model require
measurements of the conductivity of the membrane in contact with
electrolyte solutions of different concentration.

Looking at the profiles obtained by photometration, we assumed
the three layer electrochemical model as a reasonable approximation
to the membrane structure (per half width). In this model:
- the "near-surface" layer (I) contains the highest and almost
  constant amount of PAA along the layer,
- the internal layer (III) contains the lowest - and again - almost
  constant amount of PAA across the membrane,
- and the third, intermediate, layer (II) is the one in which PAA
  forms the real concentration gradient.
In each layer three phases exist:
- the solid polypropylene,
- the gel of polyacid appearing in the form of a gel of dissociated
  potassium polyacrylate with a certain amount of electrolyte sorbed
  from the ambient solution by Donnan sorption,
- and the external solution filling the volume elements free from

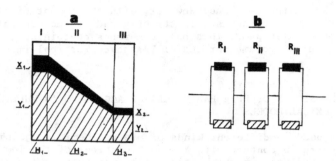

Fig. 2. Assumed three-layer model of PP-g-PAA membranes,
(a) geometric model, (b) equivalent electrical circuit.
▢ – inert polymer (PP), ▨ – gel phase (PAA),
■ – nonselective domains (voids or pores filled
with KOH solution).

polymers; this phase seems to exist in a small fraction; however
it cannot be neglected when the conductivity is considered.

The geometrical model that represents the membrane is presented in Fig. 2. The symbols x, y and h are the parameters which
describe the composition of a particular layer and the membrane as
a whole. We also assumed that a fraction of free volumes within
the membrane is proportional to the amount of the gel phase.

The corresponding equations for the resistance of each layer
are as follows:

$$R_I = \frac{h_1}{x_1 \kappa_e + y_1 \kappa_{pe}} \, , \tag{1}$$

$$R_{II} = \int_{h_1}^{1-h_3} \frac{dh}{x\kappa_e + y\kappa_{pe}} = \frac{1 - h_1 - h_3}{(y_1 - y_2)(K\kappa_e + \kappa_{pe})} \ln \frac{y_1}{y_2} \, , \tag{2}$$

$$R_{III} = \frac{h_3}{x_2 \kappa_e + y_2 \kappa_{pe}} \, , \tag{3}$$

where κ_{pe} is gel conductivity,

κ_e - conductivity of sorbed electrolyte in voids   and
K is a constant equal to x_i/y_i.

For the first and third layers the equations are rather simple
whereas for the second layer, with the gradient concentration of PAA,
the equation is more complex as it takes into account the continuous
linear variation of the polyacid content across the layer.

Summation over the resistances leads to the final equations
for the specific conductivity of the membranes:

$$R_m = 2(R_I + R_{II} + R_{III}) , \qquad (4)$$

and

$$\kappa_m = R_m^{-1} = \frac{y_1 y_2 (y_1 - y_2)(K\kappa_e + \kappa_{pe})}{2[y_2(y_1 - y_2)h_1 + y_1(y_1 - y_2)h_3 + y_1 y_2(1 - h_1 - h_3)\ln(y_1/y_2)]} . \qquad (5)$$

The conductivity curves were determined for the membranes in
equilibrium with the solutions of potassium hydroxide whose concentration ranged from 0.1 up to 2.5 mole/dm^3. These curves are
presented in Fig. 3. The geometric parameters x_i, y_i, h_i, K, and
the conductivity of the gel phase were calculated by fitting the
experimental conductivity data by Eq. (5). The standard IBM Share
Programm (no 1428) adapted to a RIAD R-32 computer was applied.

Apart from that model we also verified a number of others -
- particularly the bilayer model and the three layer model with
a constant amount of PAA in the second layer - but none of them
gave better results than the model presented.

The results of calculations can be seen in Fig. 4. Table 2
presents data on the volume fractions of the gel phase and polypropylene in each layer and in the entire membrane. The calculations were performed for three membranes of different PAA concentrations. Let us now discuss the data for the membrane containing
26% of PAA (in the dry state). In the first layer the volume fractions of the components are 0.50 of PAA and 0.45 of PP, in the next
0.38 and 0.60, and in the central layer they are 0.26 and 0.72 respectively. Altogether the volume fraction of the gel polyacid
in the swollen membrane is 0.33.

As we expected, the fraction of voids and pores is not high,
e.g. in the same membrane it is about 0.03.

The similarity of both profiles - i.e. the experimental profile
(optical, Fig. 1) and the calculated one which can be called

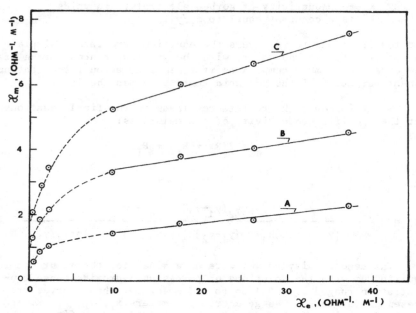

Fig. 3.   Conductivity of two-component PP-g-PAA membranes plotted
against the conductivity of ambient solution (potassium
hydroxide). Solid line calculated according to Eq. (5).
Volume fraction of PAA in dry membrane: A - 26%, B - 36%,
C - 48%

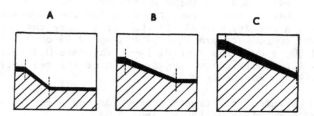

Fig. 4.   Composition of two-component PP-g-PAA membranes calculated according to electrochemical model. Volume fraction of PAA in dry membrane: A - 26%, B - 36%, C - 48%

Table 2. Composition of PP-g-PAA membranes calculated according to the three-layer model. Volume fractions of gel phase (V_{pe}) and inert polymer (V_i) in membranes swollen in potassium hydroxide solutions

Volume fractions of components (swollen state)	V_{pe}	V_i	V_{pe}	V_i	V_{pe}	V_i
membrane (total)	0.33	0.64	0.48	0.48	0.65	0.29
Ist layer	0.50	0.45	0.63	0.31	0.83	0.06
IInd layer	0.38	0.60	0.50	0.46	0.64	0.32
IIIrd layer	0.26	0.72	0.37	0.59	0.45	0.49
Volume fraction of PAA in dry membrane	0.26		0.36		0.48	

"electrochemical" (Fig. 4) – confirms that the elaborated model
represents the composition and structure with acceptable accuracy.
The model and calculations supply quantitative information about
the composition of the membranes in addition to the optical
profiles.

REFERENCES

1. M. Pegoraro, Pure Appl.Chem., 43:291 (1975).
2. A. Narębska, G. Lemańska, B. Ostrowska, Polimery, 19:374 (1974).
3. A. Narębska, R. Wódzki, in "Polymer Blends" ed. E. Martuscelli,
      R. Palumbo, M. Kryszewski, Plenum Press, New York, London
      (1980), pp. 451–467.

"MULTICOMPONENT" SEGMENTED POLYESTER SYSTEMS CONTAINING MESOGENIC

RESIDUES

Emo Chiellini,[*] Giancarlo Galli,[+] Robert W. Lenz,[++]
and Christopher K. Ober[++]

Istituto di Chimica Generale, Facoltà di Ingegneria
[+]Istituto di Chimica Organica Industriale
Università di Pisa, 56100 Pisa, Italy
[++]Materials Research Laboratory, University of
Massachusetts, Amherst, MA 01003

INTRODUCTION

Block-like and segmented polymers represent chemically bound "multicomponent" systems that, in our opinion, are able to mimic some of the non-bonded interactions occurring in blends of either compatible or not compatible polymers, and to describe phenomena connected with phase segregation and the onset of peculiar micromorphological properties. Additionally, liquid crystal polymers, even though "monocomponent" from a macrochemical point of view in that constituted of only one polymeric material, in reality do behave, under certain selected thermodynamic conditions, as mechanical mixtures of at least two components.

These premises can be taken as the major reasons for the rationale that brought us to contribute to meetings, such as the first[1] and the second Joint Polish-Italian Seminar, expressely devoted to "polymer blends". In this context, we present the results of the work done in part at the University of Pisa and in part at the University of Massachusetts in the field of the preparation and characterization of polyester systems containing *p*-oxybenzoate residues, as the anisotropic moieties, embedded in structurally ordered macromolecular segments. Linking units in the *hard* segments are the terephthaloyl (Series I, IV, and V), oxalyl (Series II) and carbonyl

(Series III) residues, respectively:

$$-\overset{\text{O}}{\underset{\text{O}}{\text{C}}}-\hspace{-2pt}\bigcirc\hspace{-2pt}-\text{O}-\text{X}-\text{O}-\hspace{-2pt}\bigcirc\hspace{-2pt}-\overset{\text{O}}{\underset{\text{O}}{\text{C}}}-\text{O}\hspace{-2pt}\left(\text{CH}_2\right)_m\hspace{-2pt}\text{O}-$$

Series I   X = $-\overset{\text{O}}{\underset{\text{O}}{\text{C}}}-\bigcirc-\overset{\text{O}}{\underset{\text{O}}{\text{C}}}-$;   **1** (m = 2-10)

Series II   X = $-\overset{\text{O}}{\underset{\text{O}}{\text{C}}}-\overset{\text{O}}{\underset{\text{O}}{\text{C}}}-$;   **2** (m = 2-12)

Series III X = $-\overset{\text{O}}{\underset{\text{O}}{\text{C}}}-$;   **3** (m = 2-12)

$$-\overset{\text{O}}{\underset{\text{O}}{\text{C}}}-\hspace{-2pt}\bigcirc\hspace{-2pt}-\text{O}-\overset{\text{O}}{\underset{\text{O}}{\text{C}}}-\hspace{-2pt}\bigcirc\hspace{-2pt}-\overset{\text{O}}{\underset{\text{O}}{\text{C}}}-\text{O}-\hspace{-2pt}\bigcirc\hspace{-2pt}-\overset{\text{O}}{\underset{\text{O}}{\text{C}}}-\text{O}\left(\text{CH}_2-\overset{*}{\underset{\text{R}}{\text{CH}}}-\text{O}\right)_n$$

Series IV   R = H ;   **4** (n = 1-22)

Series V   R = CH_3; **5** (n = 1-20)

The conformationally *soft* portion of the segmented polyester is represented by alkyl or oxyalkyl residues having different length and structure. In the case of ether segments, systems characterized by a prevailing chirality have been also interestingly used.

Thermooptical properties of all materials are reported along with a description of thermotropic mesophases generated under heating and cooling cycles. Considerations on the structure-property relationship are put forward, being aware that they may stimulate the interest of people keen on conventional multicomponent polymer systems.

SYNTHESIS OF POLYMERS

The polymers of Series I-III were prepared by step-wise poly-condensation of α,ω-bis(4-hydroxybenzoyloxy)alkane derivatives with the appropriate diacid chloride.[2-4] While samples **1** were obtained by reaction with terephthaloyl chloride in 1,1,2,2-tetrachloroethane samples **2** and **3** were prepared in dioxane with oxalyl chloride and phosgene, respectively:

In all cases the runs were carried out at room temperature using dry pyridine as the HCl acceptor. All polymers precipitated out of the reaction mixture and the polymerization proceeded in heterogeneous phase.

The polymers of Series IV-V were obtained by reaction of the diacid chloride of bis(4-carboxyphenyl)terephthalate with α,ω-glycols or glycolethers in 1,2-dichloroethane at 60°C with excess pyridine:[5,6]

While in the case of Series IV all the polymers remained soluble in the polymerization medium even after cooling to room temperature, samples of Series V precipitated since the early reaction stages.

In all cases linear, structurally ordered polymers were obtained with an alternating placement of hard and soft segments. However, according to the former procedure the mesogen-type unit consisting of two oxybenzoyl residues interconnected by a relatively rigid and polarizable group (i.e., terephthalate, oxalate, and carbonate) was built in during the polymerization, whereas in the latter case it was formed prior to polycondensation. It has to be stressed that in neither istances the monomers themselves or any precursors were liquid crystalline.

The polymeric products had medium molecular weights, as evidenced by the values of their viscosity in the range 0.1-0.3 dl/g in various solvents at 25-45°C. Oligomers with values of $\overline{M}_w/\overline{M}_n = 1.2-1.7$ were obtained in the case of oxalates **2**. Polymers **5** containing chiral centers in the spacer units were optically active and the sign of the optical rotation at 578 nm was positive as that of the starting glycolethers.[6,7]

LIQUID CRYSTAL PROPERTIES

All the polymers in Series I displayed thermotropic liquid crystal properties (Table 1)[2] Both melting and clearing (transition to isotropic melt) temperatures, T_m and T_i respectively, markedly depended on the length of the spacer and showed a regular trend with a zig-zag decrease for polymers containing up to eight methylene units in the soft segment (Figure 1)[8] The temperature range of stability of the mesophase was generally broader for the polymers with an odd m value, as compared to that of samples with an even number of methylene groupings. Such an alternation especially observed for nematic → isotropic liquid transitions has been extensively investigated in low molecular weight mesogens and interpreted in terms of changes in either anisotropic molecular polarizability or trans-gauche conformational arrangement as affected by the lengthening of the spacer[9]   Odd-even effects are recognized also in homologous series of polymers[10] and have been recently related to an anisotropic orientation of the flexible part within the liquid crystal phase[11] The increase of T_i with further lengthening of the spacer ($m>8$) in the present series has to be most likely attributed to the occurrence of a thermotropic smectic mesophase[2,12]

Table 1. Physical Properties of Series I Polymers[2,8]

Sample	m	η_{inh}[a] (dl/g)	T_m (°C)	T_i (°C)	ΔT (°C)
1_2	2	0.18	342	365[b]	23
1_3	3	0.26	240	315[b]	75
1_4	4	0.23	285	345[b]	60
1_5	5	0.14	175[c]	267	92
1_6	6	0.28	227	275	48
1_7	7	0.20	176	253	77
1_8	8	0.20	197	220	23
1_9	9	0.27	174	233	59
1_{10}	10	0.34	220	267	47

[a] In p-chlorophenol at 45°C. [b] Observed visually, not by DSC. [c] Flow temperature, no melting endotherm.

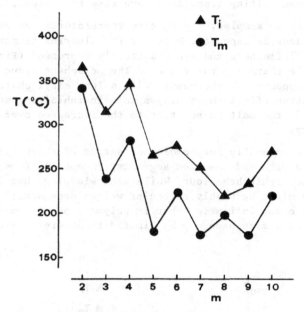

Fig. 1. Dependence of Series I polymer -●- melting temperature (T_m) and -▲- clearing temperature (T_i) on spacer length m.

The values of the thermodynamic parameters for the clearing transition, ΔH_i and ΔS_i, were lower than the corresponding values for the melting transition and generally increased with spacer length. The increase of ΔS_i with m indicates that, for polymer molecules with longer aliphatic spacers, the conformations of these chains and the placement of the mesogenic units were increasingly more ordered for the mesophase (smectic) relative to the isotropic phase[2].

An even more pronounced effect of the flexible spacer on ultimate thermal and liquid crystal properties was encountered in samples of Series II. The polymers with an even number of methylene units in the aliphatic segments, $m = 4$ through 10, were found to be mesomorphic by combined DSC analysis and observation on the hot-stage of a polarizing microscope[3]. Schlieren textures[13] with broad, diffuse extinction bands were generally observed. Sample 2_2 ($m = 2$) had too a high melting point to be fully investigated, whereas sample 2_{12} ($m = 12$)

after a broad melting transition gave rise to a clear, isotropic melt.

For these samples, the melting temperatures decrease on lengthening the spacer, as expected. However, the clearing temperatures appear to be more influenced and are drastically depressed (Figure 2). As a result the thermal persistence of the mesophase reduces and eventually disappears for the sample with a long alkyl chain ($m = 12$), in which dilution effects must play a role in inhibiting an ordered arrangement in the melt in addition to the increased overall chain flexibility.

Quite unusually for homologous series of thermotropic polymers, the samples with odd-numbered segments, except 2_3 ($m = 3$), did not show a mesomorphic behaviour. While acknowledging that liquid crystal properties might be highly molecular weight dependent,[14],[15] this appears as a clear indication that in polymer, or oligomer, systems the onset of a mesophase may be limited to discrete values of m.

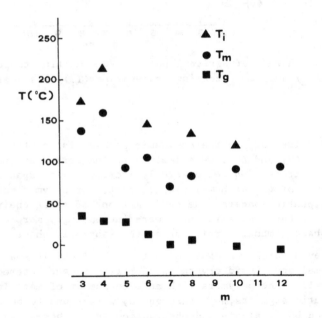

Fig. 2. Dependence of Series II polymer -●- melting temperature
(T_m), -▲- clearing temperature (T_i), and -■- glass transition temperature (T_g) on spacer length m.

Such conclusion was recently inferred by investigation of the mesomorphic properties of polyesters of p,p'-bibenzoic acid, on account
of the high dependence of the mesophase stability upon spacer length
and parity[16] In this context, we have to stress that bis(4-carboetoxy-
phenyl)oxalate and bis(4-carbohexoxyphenyl)oxalate, taken as structural model compounds, melt at 128 and 79°C respectively, without
displaying any liquid crystalline characteristics. These results
substantiate the existence of definite chain effects for the establishment of mesomorphic behaviour in the present series of samples,
in that sequences of rigid aromatic units flanked by flexible segments
are only capable of imparting peculiar liquid crystal properties.

Finally, even though the samples of Series II are of great
interest from a speculative standpoint, their practical application
seems somewhat dubious at the present stage of our investigations.
In fact, all the samples were found to exhibit low thermal stability
and yielded, after storage at room temperature, undefined products
that had lost almost completely their mesophasic properties. Such
behaviour may be related to the well known lability of poly- and
oligo(oxalate)s involving decomposition and depolymerization[17] Al-
ternatively, crystallization-induced reorganization might have occurred through interchange reactions[18] leading to more crystalline,
not mesophasic samples.

A completely different thermal behaviour was exhibited by po-
ly(carbonate)s of Series III[4] All the samples displayed typical
features of semicrystalline polymers, but in no case unambiguous
evidence of liquid crystal properties, even monotropic, could be
observed. The DSC curves showed multiple endothermic transitions
preceded by an intense pre-melt crystallization exotherm. Such phenomena may be due to the occurrence of solid state polymorphism,
analogous to a wide variety of mesomorphic polymers[19,20] Sample 3_3
($m = 3$) displayed, after melting, a weak birefringence in regions of
melt flow probably due to some shear orientation. Melting temperatures, and accordingly glass transition temperatures, clearly decrease in an alternating fashion as the series is ascended, in going
from $m = 2$ up to $m = 12$ (Figure 3)[4]

The lack of liquid crystal properties in this polymer series
was not completely unexpected and inspection of molecular models
allows one to observe only non-linear, highly bent conformations of
the diphenyl carbonate repeating unit. Very recently, investigations
by Flory and coworkers[21] on model analogues of polycarbonates have
shown that in the crystal the carbonate group is coplanar and with

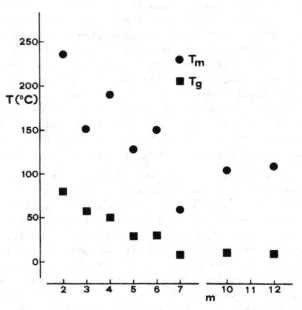

Fig. 3. Dependence of Series III polymer -●- melting temperature
(T_m) and -■- glass transition temperature (T_g) on spacer
length m.

both phenylene groups distorted by a 45° angle.

With the aim of better elucidating the role played by the flexible spacer in the present polyester systems, polymer Series IV and
V were synthesized[5,6] They are constituted by the same mesogenic aromatic triad and various glycolethers characterized by either a well
defined structure ($n = 1-4$) or an oligomeric nature ($DP_n = 4.1-22.3$).

Interestingly, the samples of Series IV with spacers consisting
of ethyleneglycol with $n = 2$, 3, 4 and with $\overline{DP}_n = 4.1$ showed two
liquid crystal transitions, T_{LC} (liquid crystal-to-liquid crystal
transition) and T_i, that could be revealed by DSC and thermal-optical
analysis (TOA) (Table 2). For polymer $4_{8.1}(\overline{DP}_n = 8.1)$ there was apparently only one mesophase, whereas on further lengthening of the
flexible segment the liquid crystallinity vanished. Polymers with a
spacer of $\overline{DP}_n > 10$ exhibited only a faint birefringence on crystalli-

Table 2. Physical Properties of Series IV Polymers.[5]

Sample	n (or $\overline{DP}_n$)	$\lvert \eta \rvert$[a] (dl/g)	T_m (°C)	T_{LC} (°C)	T_i (°C)
4_1	1.0	0.18	342	–	365
4_2	2.0	0.23	185	222	288
4_3	3.0	0.25	180	203	257
4_4	4.0	0.21	121	211	245
$4_{4.1}$	4.1	0.28[b]	103	158	195
$4_{8.7}$	8.7	0.26[b]	102	–	242
$4_{13.2}$	13.2	0.12[b]	91	–	–
$4_{22.3}$	22.3	0.11[b]	85	–	–

[a] In trifluoroacetic acid at 25°C. [b] In 1,2-dichloro-ethane at 25°C.

zation and no mesophase behaviour. The melting temperatures drastically decreased in going from sample 4_1 ($n = 1$) to 4_4 ($n = 4$), while for samples with $n > 4$ it was close to 100°C or even lower. The isotropization temperatures did not show the same great depression as the melting temperatures, but rather decreased more gradually. The thermal stability of mesomorphic properties was consistently higher for the last members in the series (Figure 4).

By comparing the results obtained for samples based on spacers of well defined structure with those for samples of Series I with alkyl spacers having the same number of atoms, it appears that the melting and clearing points are roughly the same. This observation seems to support previous suggestions that an oxygen atom is stereochemically equivalent to a methylene group when inserted in flexible spacers[22] Another interesting feature observed in Series IV is the difference between polymers having segments of well defined composition compared to those constituted by a distribution of oligomers. For istance , sample $4_{4.1}$ ($\overline{DP}_n = 4.1$) had both a lower T_m and a much narrower liquid crystal range than sample 4_4 ($n = 4$). This result may be attributed to a low compatibility of spacer blocks of different lengths analogous to polyesters containing various polymethylene glycols in a random distribution.[1,2]

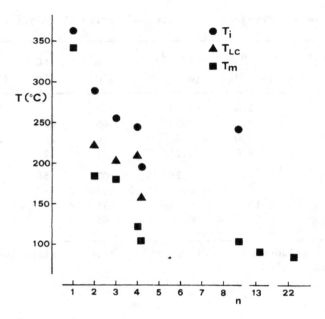

Fig. 4. Dependence of Series IV polymer -■- melting temperature
(T_m), -▲- liquid crystal-to-liquid crystal transition temperature (T_{LC}), and -●- clearing temperature (T_i) on spacer
length n.

Optical observations of mesophase textures have been previously
discussed in detail[5] It is to be stressed here that they are generally in excellent agreement with thermal analysis data and indicate
the existence of smectic and nematic phases in samples 4_2 ($n = 2$)
through $4_{4.1}$ ($\overline{DP}_n = 4.1$). On the contrary, polymer $4_{8.7}$ ($\overline{DP}_n = 8.7$)
displayed a weakly birefringent, possibly nematic, texture.

Indeed, the occurrence of liquid crystal polymorphism in the
present series does not seem unlikely, in consideration of the lath-
like character of the mesogenic core and of the nature and length of
the spacer[23] Furthermore, the model compounds bis(4-carboethoxy-
phenyl)terephthalate and bis(4-carbodecoxyphenyl)terephthalate exhibit
smectic-nematic[5] and smectic-smectic[24] dimorphism, respectively.

In order to establish whether any chemical and stereochemical
control could be exerted on the polymer liquid crystal properties

by the spacer, Series V polymers containing chiral centers in the ether segments were also synthesized (Table 3).[5,6]

The samples containing racemic spacers with $\overline{DP}_n$ less than approximately ten were found unambiguously to have mesomorphic properties by means of their optical behaviour and stir-opalescence. No melting transition could be identified by DSC, according to a mainly glassy nature, while the flow temperature was detected by optical analysis. No visual clearing was oserved and these polymers showed no isotropization endotherm below decomposition[5]

On the contrary, polymers derived from optically active glycol-ethers revealed by DSC distinct melting endotherms and, except 5_2. ($n = 2$), clearing peaks[6] Both transitions had on cooling rather high degrees of supercooling (20–50°C) and this is rather unusual for isotropic → mesophase transitions that are commonly known to exhibit very little or no supercooling. It is worth noting that the sample derived from (S)-1,2-propanediol has a much higher melting temperature and a lower mesophase stability than that of the corresponding polymer derived from the racemic diol. Apparently, the configurational regularity of the chiral segment has a pronounced effect in enhancing the propensity of the polymer to crystallize, with a consequent diminishing of its capability to form stable mesophases.

Table 3. Physical Properties of Series V Polymers[5,6]

Sample	n (or $\overline{DP}_n$)	Absolute[a] Configuration	$\lvert\eta\rvert^{25[b]}$ (dl/g)	$\lvert\alpha\rvert^{25}_{578}$[c]	T_m (°C)	T_{LC} (°C)	T_i (°C)
5_1	1.0	(R)(S)	0.12	–	262	–	nd
5_1.	1.0	(S)	$0.02^{d)}$	+ 9.5	334	–	360
5_2.	2.0	(S)	0.20	$+17.2\,^{e)}$	130	286	>320
5_3.	3.0	(S)	$0.03^{d)}$	+20.0	275	305	321
$5_{6.6}$	6.6	(R)(S)	0.08	–	181	–	nd
$5_{20.3}$	20.3	(R)(S)	$-\,^{f)}$	–	188	–	–

a) Absolute configuration of the propyleneglycol repeating unit.
b) In 1,2-dichloroethane, if not otherwise indicated. c) In fuming H_2SO_4. d) Inherent viscosity in fuming H_2SO_4 at 25°C.
e) $\lvert\alpha\rvert^{25}$ +26.6 in $CHCl_3$. f) Insoluble.

Nevertheless, the samples in this series show melting temperatures
that are very dependent on the length of the optically active ether
segment, in spite of their identical chirality and optical purity.
By contrast, no marked effect on clearing temperatures is observed.
Consequently, sample 5_2• ($n=2$) possesses a liquid crystal state
expanding over an extremely wide thermal interval (~200°C).

Another particularly interesting features of optically active
polyesters **5** is the occurrence of selective reflection of visible
light. In fact, sample 5_2• ($n=2$) in the range 190–240°C, and sample
5_3• ($n=3$) in the range 270–290°C show an orange–red and a blue–green
reflection, respectively. This demonstrates clearly the existence
of a twisted chiral structure either smectic or nematic (cholesteric). Unfortunately, neither samples could be fully investigated
by combined DSC and polarizing microscope analyses because of the
high temperatures involved. However, thermal and textural features
of sample 5_2• ($n=2$) strongly suggest the presence of a twisted
smectic phase (130–286°C) and a cholesteric phase (T > 286°C) (Figure
5). A similar liquid crystal dimorphism appears to exist also in
sample 5_3• ($n=3$).[6]

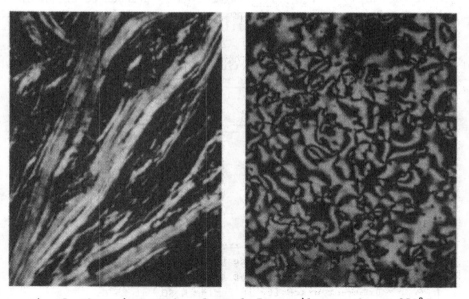

Fig. 5. Photomicrographs of sample 5_2•: oily streaks at 250°C
(left), and schlieren texture at 295°C (right). Cross
polars, original magnification 250×.

This interpretation is further supported by the behaviour shown by the model compound bis|4-carbo-(2'-methylbutoxy)phenyl)|terephthalate, for which the onset of a smectic and a cholesteric phase was detected at 135 and 146°C, respectively[6].

To our knowledge, this is the first example of the coexistence of both twisted smectic and cholesteric phases in thermotropic liquid crystal polymers. Previous preparations of thermotropic polymers by the use of chiral derivatives both incorporated in the macromolecular backbone and pendant to it as side chain substituents (comb-like polymers) resulted in either cholesteric[25-28] or smectic[27-29] polymeric products.

CONCLUSIONS

One major conclusion that can be drawn from the presented results is related to the effect of the molecular structure of the rigid, anisotropic core of the polymers under investigation on their thermotropic behaviour. Thus, if one takes both the mesophasic-to-isotropic liquid transition temperature and the temperature range of persistence of such mesophase as a qualitative index of the effectiveness of the central group X in imparting liquid crystal properties to the polymer system, the following order can be established:

It is evident that bridging units that serve to expand the anisotropic polarizability, while preserving the rigidity and linearity, of the rod-like core are most suitable in promoting a mesomorphic behaviour.[9]

Furthermore, the onset of liquid crystallinity in the melt is markedly affected even by the structure of the flexible spacer. Originally, such segments were introduced in the main chain of thermotropic polymers mainly with the aim of lowering melting, and consequently clearing, temperature.[30] Beside that, it is now clear that the structure and conformational array of the spacer are of relevance to either favouring or hindering thermotropic mesomorphism. In this respect, oxygen-containing units, when compared to equivalent methylene segments, are more effective in inducing smectic-like order in the melt, while not affecting substantially the thermal stability of the mesophasic state. Such liquid crystal behaviour

can be preserved in samples in which the amount of glycolether units
as a fraction of the whole reaches values as high as 55-60 wt-%.
Moreover, appropriate optically active spacers can give rise to
chiral twisted mesophases, whose intrinsic properties such as reflection of circularly polarized light in the visible region are temperature dependent.

From the above stressed points it comes out rather evidently how
the micromorphological properties in the melt of structurally defined
segmented polyesters containing the same mesogen-type moiety are quite
markedly influenced even by subtle differences in both the hard and
soft portion of the repeating units. These findings are in our
opinion worth of attention, in that they highlight the potential of
the investigated polymers for both applicative and speculative stand-
points in the field of multicomponent polymeric systems. In fact,
their well defined chemical and stereochemical structure provides,
from one side plain monocomponent systems with some peculiar characteristics of the multicomponent-type, and from the other the opportu-
nity of evaluating, as a further prosecution of the present work,
their capability as phase inducers in suitable polymer blends.

ACKNOWLEDGEMENTS

The authors wish to thank the Italian National Research Council
(CNR) for partial support of the work.   NATO is also gratefully
acknowledged for providing the grant RG 076.81.

REFERENCES

1. E.Chiellini, R.W.Lenz, and C.Ober, Multicomponent polyester
   systems with mesogenic units, in "Polymer Blends", E.Martuscelli,
   R.Palumbo, and M.Kryszewski, eds., Plenum Press, New York (1980).
2. C.Ober, J.I.Jin, and R.W.Lenz, Thermotropic polyesters with either
   dyad or triad aromatic ester mesogenic units and flexible polymethylene spacers in the main chain, Polymer J. 14:9 (1982).
3. G.Galli, P.Nieri, C.K.Ober, and E.Chiellini, Preparation and
   properties of mesomorphic oligo(oxalate)s, Makromol.Chem., Rapid
   Commun. 3:543 (1982).
4. P.Nieri, Thesis, University of Pisa (1981).
5. G.Galli, E.Chiellini, C.K.Ober, and R.W.Lenz, Structurally ordered
   thermotropic polyesters of glycol ethers, Makromol.Chem.
   183:2693 (1982).
6. E.Chiellini, G.Galli, C.Malanga, and N.Spassky, Structurally
   ordered thermotropic polyesters of optically active propyleneglycol
   ethers, Polymer Bull. submitted.

7. C.Malanga, N.Spassky, R.Menicagli, and E.Chiellini, Synthesis of optically active oligoethers of (S)-1,2-propanediol, Polymer Bull. submitted.

8. C.K.Ober, Thesis, University of Massachusetts (1982).

9. G.W.Gray, Liquid Crystals and molecular structure: Nematics and cholesterics, in: "The Molecular Physics of Liquid Crystals", G.R.Luckurst and G.W.Gray,eds., Academic Press, New York (1979).

10. A.C.Griffin and S.J.Havens, Thermal properties and synthesis of three homologous series of thermotropic liquid crystalline backbone polyesters, J.Polym.Sci., Polym.Phys.Ed. 19:951 (1981).

11. A.Roviello and A.Sirigu, Odd-even effects in polymeric liquid crystals, Makromol.Chem. 183:895 (1982).

12. V.Frosini, A.Marchetti, and S.De Petris, Low temperature phase transition in some thermotropic polyesters quenched from the liquid crystalline state, Makromol.Chem., Rapid Commun. 3:795 (1982).

13. D.Demus and L.Richter, "Textures of Liquid Crystals", Verlag Chemie, Weinheim (1978).

14. A.Blumstein, S.Vilasagar, S.Ponrathnam, S.B.Clough, and R.B. Blumstein, Nematic and cholesteric thermotropic polyesters with azoxybenzene mesogenic units and flexible spacers in the main chain, J.Polym.Sci., Polym.Phys.Ed. 20:877 (1982).

15. M.Portugall, H.Ringsdorf, and R.Zentel, Synthesis and phase behaviour of liquid crystalline polyacrylates, Makromol.Chem. 183:3211 (1982).

16. W.R.Krigbaum, J.Asrar, H.Toriumi, A.Ciferri, and J.Preston, Aromatic polyesters forming thermotropic smectic mesophases, J.Polym.Sci., Polym.Lett.Ed. 20:109 (1982).

17. A.Alksnis and Dz.Deme, Investigations on equilibrium of poly-(ethylene oxalate) and 1,4-dioxane-2,3-dione, J.Polym.Sci., Polym.Chem.Ed. 17:2701 (1979).

18. R.W.Lenz and A.N.Schuler, Crystallization-induced reactions of copolymers. Important reaction variables in the reorganization of random to block copolymers, J.Polym.Sci., Polym.Symp. 63:343 (1978).

19. S.Antoun, R.W.Lenz, and J.I.Jin, Thermotropic polyesters with flexible spacers in the main chain, J.Polym.Sci., Polym.Chem.Ed. 19:1901 (1981).

20. A.Roviello and A.Sirigu, Solid and liquid crystalline phases of polyalkanoates containing the 1,4-phenylene-(2-methylvinylene)-1,4-phenylene group, Preprints IUPAC Symposium on Macromolecules, Florence, 3:290 (1980).

21. B.Erman, D.C.Marvin, P.A.Irvine, and P.J.Flory, Optical aniso-
    tropies of model analogues of polycarbonates, Macromolecules
    15:664 (1982).
22. P.Meurisse, C.Noel, L.Monnerie, and B.Fayolle, Polymers with
    mesogenic elements and flexible spacers in the main chain:
    Aromatic-aliphatic polyesters, Br.Polym.J. 13:55 (1981).
23. G.W.Gray, Liquid crystals and molecular structure: Smectics,
    in: "The Molecular Physics of Liquid Crystals", G.R.Luckurst
    and G.W.Gray, eds., Academic Press, New York (1979).
24. G.Galli and E.Chiellini, unpublished results.
25. S.Vilasagar and A.Blumstein, Cholesteric thermotropic polymers
    with mesogenic moieties and flexible spacers in the main chain,
    Mol.Cryst.Liq.Cryst.Lett. 56:263 (1980).
26. H.Finkelmann, J.Koldehoff, and H.Ringsdorf, Synthesis and charac-
    terization of liquid crystalline polymers with cholesteric
    phase, Angew.Chem.Int.Ed.Engl. 17:935 (1978).
27. H.Finkelmann, H.Ringsdorf, W.Siol, and J.H.Wendorff, Synthesis
    of cholesteric liquid crystalline polymers, Makromol.Chem.
    179:829 (1978).
28. V.P.Shibaev, N.A.Plate, and Ya.S.Freidzon, Thermotropic liquid
    crystalline polymers. Cholesterol containing polymers and co-
    polymers, J.Polym.Sci., Polym.Chem.Ed. 17:1655 (1979).
29. V.P.Shibaev, Ya.S.Freidzon, and N.A.Plate, Liquid crystal cho-
    lesterol-containing polymers, Dokl.Akad.Nauk SSSR 227:1412
    (1976).
30. A.Roviello and A.Sirigu, Mesophasic structures in polymers.
    A preliminary account on the mesophases of some polyalkanoates
    of p,p'-di-hydroxy-α,α'-di-methyl benzalazine, J.Polym.Sci.,
    Polym.Lett.Ed. 13:455 (1975).

AUTHOR INDEX

A

Aliverti, M., 205
Avella, M., 193

B

Bartczak, Z., 243
Bianchi, L., 57
Bukowski, Z., 257

C

Cammarata, E., 205
Chiellini, E., 267

D

Danesi, S., 34
Dobkowski, Z., 85, 157
D'Orazio, K., 111, 127

G

Galeski, A., 223
Galli, G., 267
Greco, R., 111, 127

J

Jeszka, J.K., 165

K

Kammer, H.W., 19
Kowalewski, T., 223
Kozłowski, M., 101
Krajewski, B., 157
Kryszewski, M., 165, 173, 223, 243,
Kuczyński, J., 179

L

Lanzetta, N., 193
Lenz, R.W., 267

M

Maglio, G., 41, 193
Malinconico, M., 193
Mancarella, C., 111
Martuscelli, E., 57, 73, 111, 127
Musto, P., 193
Myśliński, P., 157

N

Narębska, A., 257

O

Ober, C.K., 267

SUBJECT INDEX